Tourism and Poverty Alleviation in Nature Conservation Areas

This timely book delves into the intricate relationship between tourism and poverty with a specific focus on nature conservation areas, using case studies of island economies in a developed country, Japan, and a developing one, Vietnam.

The volume asserts that although the concept of pro-poor tourism has often linked tourism with poverty, limited research has examined this link from diverse perspectives, including those of developed and developing countries where poverty can understand in absolute or relative forms. Notably, the book considers the voices of local residents in these areas, particularly the impoverished living in tourist destinations in Vietnam. This is essential for influencing conservation efforts and making poverty alleviation more achievable. Readers, therefore, gain an understanding of why tourism and poverty alleviation are crucial for every economy within the context of nature conservation areas.

This volume is a pivotal resource for scholars in tourism, particularly those focused on teaching and researching tourism geographies and sustainable development. It holds particular significance for scholars examining emerging nations across Asia.

Nguyen Van Hoang holds a BA in Tourism Geography and an MA in Natural Resources Preservation, Rational Exploitation and Restoration from Vietnam National University in Ho Chi Minh City. He obtained his second MA and PhD degrees from Hiroshima University in Japan. He is currently Lecturer at the Faculty of Tourism, Nguyen Tat Thanh University in Vietnam. His research primarily focuses on sustainable tourism development, tourism and poverty alleviation, particularly in protected areas and marine environments. He has published numerous papers in Scopus-listed journals such as Tourism Planning & Development, Tourism in Marine Environments, Parks, Tourism & Hospitality Research, Cogent Social Sciences and Current Issues in Tourism. He is passionate about teaching subjects related to sustainable tourism, ecotourism, tourism planning and tourism geography. Additionally, he co-founded the Vietnam Tourism Research Network, aimed at enhancing the publication capacity of Vietnamese researchers in the tourism field.

Routledge Insights in Tourism Series
Series Editor
Anukrati Sharma
*Head & Associate Professor of the Department of Commerce
and Management at the University of Kota India*

This series provides a forum for cutting edge insights into the latest developments
in tourism research. It offers high quality monographs and edited collections that
develop tourism analysis at both theoretical and empirical levels.

**Potentials, Challenges and Prospects of Halal Tourism Development
in Ethiopia**
Mohammed Jemal Ahmed and Atilla Akbaba

Diasporic Mobilities on Vacation
Tourism of European-Moroccans at Home
Lauren B Wagner

**Overtourism and Cruise Tourism in Emerging Destinations on the
Arabian Peninsula**
Manuela Gutberlet

Pseudo-Authenticity and Tourism
Preservation, Miniaturization, and Replication
Jesse Owen Hearns-Branaman and Lihua Chen

Developing Industrial and Mining Heritage Sites
Lavrion Technology and Cultural Park, Greece
Taşkın Deniz Yıldız

Innovation Strategies and Organizational Culture in Tourism
Concepts and Case Studies on Knowledge Sharing
Edited by Marco Valeri

Tourism and Poverty Alleviation in Nature Conservation Areas
A Comparative Study Between Japan and Vietnam
Nguyen Van Hoang

For more information about this series, please visit: *www.routledge.com/Routledge-Insights-in-Tourism-Series/book-series/RITS*

Tourism and Poverty Alleviation in Nature Conservation Areas

A Comparative Study Between
Japan and Vietnam

Nguyen Van Hoang

Routledge
Taylor & Francis Group

LONDON AND NEW YORK

First published 2024
by Routledge
4 Park Square, Milton Park, Abingdon, Oxon OX14 4RN

and by Routledge
605 Third Avenue, New York, NY 10158

Routledge is an imprint of the Taylor & Francis Group, an informa business

© 2024 Nguyen Van Hoang

British Library Cataloguing-in-Publication Data
A catalogue record for this book is available from the British Library

ISBN: 978-1-032-80412-5 (hbk)
ISBN: 978-1-032-80413-2 (pbk)
ISBN: 978-1-003-49674-8 (ebk)

DOI: 10.4324/9781003496748

Typeset in Times New Roman
by Apex CoVantage, LLC

Contents

Tables

Figures

Abbreviations

ADB	Asian Development Bank
CLCTMB	Cu Lao Cham Tourism Management Board
CPRGS	Comprehensive Poverty Reduction and Growth Strategy
GDP	Gross Domestic Product
GOV	Government of Vietnam
GSOV	General Statistics Office of Vietnam
ICEM	International Centre for Environmental Management
JNTO	Japan National Tourism Organization
JTA	Japan Tourism Agency
MARD	Ministry of Agriculture and Rural Development
MLIT	Ministry of Land, Infrastructure, Transport and Tourism
MPA	Marine Protected Area
NGO	Non-governmental Organization
NSTD	National Strategy for Tourism Development
NWHS	Natural World Heritage Site
OECD	Organization for Economic Co-operation and Development
PA	Protected Area
PPT	Pro-poor Tourism
SIDS	Small Island Developing States
THPC	Tan Hiep People's Committee
UN	United Nations
UNDP	United Nations Development Program
UNESCO	United Nations Educational, Scientific, and Cultural Organization
UNWTO	United Nations World Tourism Organization
VNAT	Vietnam National Administration of Tourism
WB	World Bank

Acknowledgments

This book is compiled on the basis of research results from the author's doctoral thesis. The author would like to express gratitude to the following individuals and institutions for their invaluable assistance:

- Hiroshima University, where the author completed his doctoral program
- Nguyen Tat Thanh University, where the author is currently working

Introduction

Poverty alleviation is a crucial goal for many countries around the world, and tourism is widely recognized as a significant contributor to economic growth and the attainment of the United Nations' Sustainable Development Goals (UN SDGs) (United Nations World Tourism Organization [UNWTO], 2011, 2018a). The pro-poor tourism (PPT) concept emerged in 1999 as one key approach to this policy link. PPT aims to "increase the net benefits for the poor from tourism, and ensure that tourism growth contributes to poverty reduction" (Ashley, Roe, & Goodwin, 2001, p. viii). The use of tourism for poverty alleviation has been promoted at three different levels: the development agency level (e.g., Netherlands Development Organization, German International Cooperation), national level (e.g., the UK Department for International Development) and international institution level (e.g., the UNWTO's Sustainable Tourism—Eliminating Poverty Initiative and Foundation) (Hummel & van der Dium, 2012).

The potential contribution of tourism to poverty alleviation has received increased attention from previous scholars (e.g., Alam & Paramati, 2016; Blake, 2008; Croes & Rivera, 2017; Truong, Liu, & Pham, 2020). This has led to a wide discussion of this relationship in many countries, particularly in developing countries where tourism is considered an important tool for poverty alleviation. The link between tourism and poverty has also been discussed in nature conservation areas (e.g., Job & Paesler, 2013; Leisher, Beukering, & Scherl, 2007; H. Manwa & Manwa, 2014). In these special environments, tourism is often viewed as an alternative livelihood for local people whose primary sources of income (e.g., fishing and/or forestry, which are dependent on natural resources) may be impacted by conservation purposes and restrictive regulations. Additionally, in some cases, tourism is seen as a means of reducing poverty by generating new income for local people (e.g., Leisher et al., 2007; Soliman, 2015).

Although poverty is often associated with many developing countries, it is a complex, multidimensional concept that also affects developed countries. Despite the multidimensionality of poverty, it can be understood as relative poverty, such as income inequality, which is faced by many developed countries like Japan, or absolute poverty, which is suffered by a majority of developing countries like Vietnam. In the case of Japan, the country has a high-income inequality and poverty rates among OECD members, ranking 16th in income inequality with a Gini coefficient

DOI: 10.4324/9781003496748-1

of 0.33 in 2015, and a poverty rate of 16.1% in 2012 (OECD, 2018). Tourism, along with other industries, is being promoted in Japan to contribute to the local economy, especially in many less-favored areas, and has a positive impact on the Japanese economy as a whole (Japan Tourism Agency, 2014, 2016, 2017).

In contrast to Japan, Vietnam has considered poverty alleviation an important task since the early 1990s. In its effort, the country has successfully reduced poverty by half between 1990 and 2000 (Government of Vietnam (GOV), 2003). Although Vietnam is no longer a least developed country (UN, 2012) and poverty rates have declined significantly in the past decade, as of 2017, 7.9% of the population continue to live under the national poverty lines of VND700,000 (US$35.0) and VND900,000 (US$45.0) per person per month in rural and urban areas, respectively (General Statistics Office of Vietnam (GSOV), 2017). According to Vietnam's Law on Tourism (GOV, 2017), tourism is officially recognized as a means for poverty alleviation, especially focused on rural and isolated areas where there is a high proportion of poor people.

While nature conservation areas emphasize conservation as a key strategy, attention is also given to the socioeconomic impacts on people living in and around them (Brockington & Wilkie, 2015). In these unique environments, tourism is often considered a conservation intervention that provides benefits to local people while contributing to conservation purposes (Sandbrook, 2010). Moreover, due to the rich biodiversity in these areas, tourism is expected to replace local primary industries, such as forestry and fishery, which have been limited by new environmental regulations (Mbaiwa & Stronza, 2010).

A number of studies have explored the benefits of protected areas to local communities and poverty alleviation through tourism (e.g., Job & Paesler, 2013; Leisher et al., 2007; H. Manwa & Manwa, 2014; Soliman, 2015; Spiteri & Nepal, 2006). Specifically, Leisher et al. (2007) revealed that marine protected areas (MPAs) can lead to better governance, health improvements, increased employment for women and poverty reduction through tourism. Meanwhile, Job and Paesler (2013) considered the links between nature-based tourism in protected areas and poverty alleviation in Kenya. Their results suggest that tourism increased the income of communities and improved their standard of living, but does not necessarily contribute to poverty reduction. However, H. Manwa and Manwa's study (2014) found that ecotourism development in forest reserves in Botswana has the potential to alleviate poverty for the local communities living nearby these protected areas. Similarly, Soliman (2015) discussed the adoption of PPT in protected areas in Egypt and indicated that tourism positively impacts the livelihoods of local people, suggesting that tourism growth together with protected areas may have a positive effect at the same time.

The role of tourism development in protected areas is widely acknowledged for improving local livelihoods economically, especially in many developing economies. However, the benefits from tourism can be distributed unequally among the local communities, leading to exclusion of the poor or leakage of benefits from the local areas to elsewhere (Sandbrook, 2010; Sekhar, 2003). Distribution of tourism benefits and the attitudes of local communities toward protected areas have been widely discussed as key barriers to conservation success (Brooks, Waylen, & Mulder, 2013; Spiteri & Nepal, 2006). Therefore, income inequality in wildlife conservation areas may lead to social conflicts, tensions among social groups and negative attitudes of local people who do not benefit directly, potentially influencing tourism development policies (Walpole & Goodwin, 2001). These phenomena suggest that nature-based tourism in protected areas is challenging in terms of maintaining local benefits while protecting natural values. Additionally, it reveals that PPT discourse pays little attention to income inequality in these special settings.

Although there are numerous books and studies on poverty in Southeast Asia, most of them only provide a general overview of the poverty in countries like Laos or Thailand (e.g., Rigg, 2005, 2018), but they did not mention the relationship between poverty and tourism. In Connell's study (2017), the author examines changes in Southeast Asia's tourism industry, including the emergence of specific niche markets like medical tourism. However, in addition to the positive contributions of tourism, the region also experiences problems of disproportionate development among countries. Recently, several studies on PPT have been conducted in Vietnam (e.g., Huynh, 2011; Redman, 2009; Truong, Hall, & Garry, 2014; Truong, Liu, & Pham, 2020), but none of them have considered the perspectives of local people in the context of nature conservation areas, especially in the MPAs. Meanwhile, there has been little attention paid to PPT in developed countries such as Japan, where poverty rates and income inequality are relatively high. Therefore, there is a need to discuss not only the actual impacts of tourism but also the local people's perceptions regarding the tourism–poverty nexus.

Furthermore, although the poverty and tourism development, particularly tourism in nature conservation areas, may vary in Japan and Vietnam, these countries are appropriate places to examine the concept of PPT in different contexts between developed and developing countries. This can help to better understand and exchange ideas in PPT discourse. Therefore, this book will specifically examine the interrelationship between tourism and poverty in two countries, Japan and Vietnam, with a particular focus on nature conservation areas. To find similarities and differences and understand how tourism influences poverty in these countries, one island from each country will be selected for the comparison.

In the case of Vietnam, Cu Lao Cham Island, an MPA, was selected to analyze the contribution of tourism to poverty reduction and to discover any possible inequalities arising from this process. The focus in Japan will be on Yakushima Island, a Natural World Heritage Site (NWHS), to examine whether inequalities in income distribution and spatial development exist as a result of tourism's impact. Furthermore, a concept of PPT for developed countries will also be created. Yakushima and Cu Lao Cham have been selected for this comparison because of their similarities in geographical features (both are nature conservation areas and islands) and socioeconomic characteristics (tourism is the main industry in both cases). While the development of tourism may differ in these two scenarios, they can be compared in order to learn valuable lessons from each other. This book aims to examine the connection between tourism and poverty alleviation using the PPT framework, shedding light on both the positive and negative impacts of tourism on poverty-related matters. While tourism has the potential to reduce poverty and promote regional development, it can also lead to income inequality. Consequently, this book will delve into and analyze the effects of tourism on these specific issues.

A purposeful sampling strategy was employed to select the study sites for this research. Yakushima Island and Cu Lao Cham Island were specifically chosen as they possess similarities that facilitate a comparison of the PPT concept in these two cases. First, both locations are designated as protected areas and are situated on islands, sharing comparable geographical characteristics. Second, tourism has become the primary industry in both sites, replacing traditional livelihoods such as forestry in Yakushima and fishery in Cu Lao Cham, after the areas were designated as nature conservation areas. Third, tourism is recognized as a significant contributor to the local people's livelihoods and the economies of both regions. Finally, each study site experiences distinct economic conditions, as mentioned earlier, and the impacts of tourism may differ between the two locations.

The selection of Japan, specifically Yakushima, as the focus of this study was driven by several factors. First, Japan is known for having one of the highest levels of income inequality among OECD countries. Second, there are high expectations for tourism, particularly in nature-based tourism within NWHS in Japan, to bring benefits to local communities in exchange for restricted access to natural resources (Krag & Prebensen, 2016). Third, Yakushima Island, which became the first NWHS site in Japan in 1993, is situated in Kagoshima Prefecture, where the poverty rate is relatively higher than the national average in Japan. Fourth, historically, forestry was a key industry on Yakushima but was abandoned in the 1980s to prioritize the conservation of its natural heritage. Although tourism has emerged as a significant industry, it has revealed some income distribution disparities among stakeholders and spatial variations among villages on the island (Kanetaka & Funck, 2012). Finally, as Yakushima is the inaugural NWHS in Japan, this research provides an opportunity to examine the 26-year (1993–2019) span of tourism development and its long-term effects. Thus, Yakushima Island serves as an ideal location to investigate the interplay between tourism, nature conservation and income distribution.

There are several reasons behind the selection of Cu Lao Cham Island and Vietnam for this study. First, Vietnam, despite experiencing significant economic growth and poverty reduction, remains a relatively impoverished country. Second, as a Vietnamese researcher, there is an advantage in exploring the relationship between tourism and poverty within my own country. Third, previous studies have been conducted in Vietnam on this subject (e.g., Huynh, 2011; Redman, 2009; Truong et al., 2014), but none of them have specifically examined the perspectives of local communities within protected areas, particularly MPAs. Fourth, prior to the establishment of the MPA in Cu Lao Cham Island, the local population faced a certain level of poverty. However, research indicates that since the promotion of tourism, there has been a notable improvement in the lives of local residents (Chu, 2014; Brown, 2011). Consequently, Cu Lao Cham Island is an appropriate location to investigate the connections between tourism, poverty and conservation outcomes.

For the aforementioned reasons, by studying two case studies, Yakushima and Cu Lao Cham, in the context of nature conservation areas on islands, this research aims to achieve its objectives while also distinguishing itself from previous PPT studies.

The author's aim is to address issues surrounding the relationship between tourism and poverty alleviation in nature conservation areas. Three main questions are posed to achieve this objective: (1) To what extent do local residents perceive the impact of tourism on poverty reduction in Cu Lao Cham? (2) How do tourism businesses in Yakushima perceive the impact of tourism on income distribution and spatial tourism development? (3) How do the perceptions of the tourism and poverty nexus in the two cases compare and contrast with each other? By examining the insights of local residents, this book intends to contribute to the concept of poverty reduction through tourism. This research will provide three main contributions: (1) expanding the understanding of tourism and poverty alleviation in different contexts, especially between developed and developing economies; (2) increasing the geographical scope of the PPT study by analyzing the topic in nature conservation areas; and (3) proposing a new PPT concept and approach to measuring poverty in various contexts.

The book is structured as follows, with seven chapters covering a range of topics related to tourism and poverty alleviation. Before delving into the detailed information in each chapter of the monograph, the introduction provides an overview of the background of the book and explains the reasons for choosing the research site. Chapter 1 sets the stage by providing background information on the geographic, social and political contexts of tourism development and poverty in Japan. In Chapter 2, the focus shifts to Vietnam, examining the relationship between tourism development and poverty in that country. Chapter 3 offers an in-depth look at the case study of Yakushima, providing an overview of the region and its economic prior to analyzing the impact of tourism on poverty alleviation. Similarly, Chapter 4 provides an overview of the case study of Cu Lao Cham, including a review of tourism development and the economic situation in the region. Chapter 5 explores the perspectives of local tourism enterprises on tourism development in the Yakushima case, while Chapter 6 reports on the perceptions of local people in Cu Lao Cham regarding the contribution of tourism to poverty alleviation. Finally, Chapter 7 offers a comparison of the results of the two case studies, examining the perceptions of local tourism enterprises on the impact of tourism on the local economy and barriers to tourism participation.

Reference list

Alam, M. S., & Paramati, S. R. (2016). The impact of tourism on income inequality in developing economies: Does Kuznets curve hypothesis exist? *Annals of Tourism Research, 61,* 111–126.

Ashley, C., Roe, D., & Goodwin, H. (2001). *Pro-poor tourism strategies: Making tourism work for the poor*. London: Pro-poor Tourism Partnership.

Blake, A. (2008). Tourism and income distribution in East Africa. *International Journal of Tourism Research, 10,* 511–524.

Brockington, D., & Wilkie, D. (2015). Protected areas and poverty. *Philosophical Transactions of the Royal Society B: Biological Sciences*, *370*, 1–6.

Brooks, J., Waylen, A. K., & Mulder, B. M. (2013). Assessing community-based conservation projects: A systematic review and multilevel analysis of attitudinal, behavioral, ecological, and economic outcomes. *Environmental Evidence*, *2*(2), 1–34.

Brown, P. (2011). Livelihood change around marine protected areas in Vietnam: A Case Study of Cu Lao Cham. In Canada Research Chair in Asian Studies (Ed.), *ChATSEA working paper no. 16*. Montreal: Universite de Montreal.

Chu, M. T. (2014). *Opportunities and challenges in the management and preservation of biodiversity value of Cu Lao Cham Biosphere Reserve*. Hoi An. Retrieved October 10, 2018, from www. khusinhquyenculaocham.com.vn. (In Vietnamese).

Connell, J. (2017). 'Timeless Charm'? Tourism and development in Southeast Asia. In *Routledge handbook of Southeast Asian development* (pp. 153–168). London: Routledge.

Croes, R., & Rivera, M. (2017). Tourism's potential to benefit the poor: A social accounting matrix model applied to Ecuador. *Tourism Economics*, *23*, 29–48.

General Statistics Office of Vietnam (GSOV). (2017). *Statistical yearbook of Vietnam*. Hanoi: Statistical.

Government of Vietnam (GOV). (2003). *Comprehensive poverty reduction and growth strategy* (CPRGS). Hanoi: Cartography Publishers.

Government of Vietnam (GOV). (2017). *Law on tourism*. Retrieved October 10, 2018, from www.vietnamtourism.gov.vn/index.php/docs/853. (In Vietnamese).

Hummel, J., & van der Dium, R. (2012). Tourism and development at work: 15 years of tourism and poverty reduction within the SNV Netherlands Development Organisation. *Journal of Sustainable Tourism*, *20*(3), 319–338.

Huynh, B. T. (2011). *The Cai Rang floating market, Vietnam: Towards PPT?* (Master's thesis, Auckland University of Technology, Auckland).

Japan Tourism Agency (JTA). (2014). *Economic impact of travel on Japan*. Ministry of Land, Infrastructure, Transport and Tourism (MLIT). Retrieved from www.mlit.go.jp/kankocho/en/siryou/toukei/kouka.html

Japan Tourism Agency (JTA). (2016). *White paper on tourism in Japan*. Ministry of Land, Infrastructure, Transport and Tourism (MLIT). Retrieved from www.mlit.go.jp/common/001141408.pdf

Japan Tourism Agency (JTA). (2017). *White paper on tourism in Japan*. Ministry of Land, Infrastructure, Transport and Tourism. Retrieved from www.mlit.go.jp/common/001211721.pdf

Job, H., & Paesler, F. (2013). Links between nature-based tourism, protected areas, poverty alleviation and crises—The example of Wasini Island (Kenya). *Journal of Outdoor Recreation and Tourism*, *1–2*, 18–28.

Kanetaka, F., & Funck, C. (2012). The development of the tourism industry in Yakushima and its spatial characteristics. *Studies in Environmental Sciences*, *6*, 65–82. (In Japanese).

Krag, C., & Prebensen, N. (2016). Domestic nature-based tourism in Japan: Spirituality, novelty and communing. In J. S. Chen (Ed.), *Advances in hospitality and leisure, Volume 12* (pp. 51–64). London: Emerald Group Publishing Limited.

Leisher, C., van Beukering, P., & Scherl, M. L. (2007). *Nature's investment bank: How marine protected areas contribute to poverty reduction*. London: The Nature Conservancy and WWF International.

Manwa, H., & Manwa, F. (2014). Poverty alleviation through pro-poor tourism: The role of Botswana forest reserves. *Sustainability*, *6*, 5697–5713.

Mbaiwa, J., & Stronza, A. (2010). The effects of tourism development on rural livelihoods in the Okavango Delta, Botswana. *Journal of Sustainable Tourism*, *18*, 635–656.

Organization for Economic Co-operation and Development (OECD). (2018). *Income inequality and poverty*. Paris: OECD. Retrieved from www.oecd.org/social/inequality-and-poverty.htm

Redman, D. (2009). *Tourism as a poverty alleviation strategy: Opportunities and barrier for creating backward economic linkages in Lang Co, Vietnam* (Master's Thesis, Massey University, Massey).

Rigg, J. (2005). Poverty and livelihoods after full-time farming: A South-East Asian view. *Asia Pacific Viewpoint, 46*(2), 173–184.

Rigg, J. (2018). Rethinking Asian poverty in a time of Asian prosperity. *Asia Pacific Viewpoint, 59*(2), 159–172.

Sandbrook, C. G. (2010). Local economic impact of different forms of nature-based tourism. *Conservation Letters, 3*(1), 21–28.

Sekhar, U. N. (2003). Local people's attitudes towards conservation and wildlife tourism around Sariska Tiger Reserve, India. *Journal of Environmental Management, 69*(4), 339–347

Soliman, M. (2015). Pro-poor tourism in protected areas—Opportunities and challenges: "The case of Fayoum, Egypt". *Anatolia—An International Journal of Tourism and Hospitality Research, 26*(1), 61–72.

Spiteri, A., & Nepal, S. (2006). Incentive-based conservation programs in developing countries: A review of some key issues and suggestions for improvements. *Environmental Management, 37*, 1–14.

Truong, V. D., Hall, C. M., & Garry, T. (2014). Tourism and poverty alleviation: Perceptions and experiences of poor people in Sapa, Vietnam. *Journal of Sustainable Tourism, 22*(7), 1071–1089.

Truong, V. D., Liu, X., & Pham, Q. (2020). To be or not to be formal? Rickshaw drivers' perspectives on tourism and poverty. *Journal of Sustainable Tourism, 28*(1), 33–50.

United Nations (UN). (2012). *The Millennium development goals report 2012*. New York: UN.

United Nations World Tourism Organization (UNWTO). (2011). *Policy and practice for global tourism*. Madrid: Author.

United Nations World Tourism Organization (UNWTO). (2018a). *Annual report 2017*. Madrid: UNWTO.

Walpole, M. J., & Goodwin, H. J. (2001). Local attitudes towards conservation and tourism around Komodo National Park, Indonesia. *Environmental Conservation, 28*, 160–166.

Part I
Setting the context

1 Tourism development and poverty in Japan

1.1 Tourism development: an overview

Tourism has a long and storied history in Japan, dating back to pre-modern times. During the Edo period (1603–1868), many people travelled on foot to visit Shinto shrines, Buddhist temples, and hot springs for both religious and health reasons (Nobukiyo, 2010). Following the Meiji Restoration in 1868, Japan opened up to the world, leading to greater freedom of travel with the abolishment of internal checkpoints and passports (Funck & Cooper, 2013). This period also saw the development of alternative means of transportation, including the construction of the first railway line connecting Yokohama and Tokyo in 1872, and the expansion of rail networks throughout Japan. In addition, the first Western-style hotel for foreign tourists was built in 1863, and the Japan Tourist Bureau, which is still one of the country's leading travel agencies today, was founded in 1912 with the aim of attracting more international visitors and foreign currency to Japan (Funck & Cooper, 2013). However, the impact of World War II led to a suspension of tourism in Japan until 1945. Figure 1.1 illustrates the location of Yakushima Island in Japan.

After World War II, tourism in Japan was redeveloped and can be divided into five subperiods. The first subperiod, from 1945 to 1963, is known as the reconstruction period. During this time, the public was not allowed to travel abroad for tourism due to insufficient foreign-exchange reserves to support such travel. The second subperiod, from 1964 to 1969, saw a relaxation of policies toward foreign travel. The final three subperiods are the beginning of mass tourism (1970s to 1980s), the beginning of alternative tourism (1980s to 2006) and the tourism-nation period (2006 to present) (Nobukiyo, 2010). Each of these subperiods has its own unique characteristics, which will be briefly described in the next section.

During the period from 1964 to 1969, the Japanese government eased its policies toward international travel, and more people began to travel abroad for tourism (Nobukiyo, 2010). This period also saw the Tokyo Olympics in 1964, which acted as a stimulus for the development of tourism infrastructure. Many tourists visited Tokyo and other parts of Japan for the Olympics, which led to the expansion of the hotel and transportation industries. During this period, the economic growth in Japan was remarkable, and the standard of living of Japanese people

DOI: 10.4324/9781003496748-3

Figure 1.1 Location of Yakushima island in Japan.

Source: National Parks of Japan (2019)

rose rapidly. As a result, tourism-related consumption also increased, and people started to participate in domestic tourism more frequently (Nobukiyo, 2010). The period from the 1970s to the 1980s was characterized by the beginning of mass tourism in Japan. The growth of the Japanese economy and the development of transportation infrastructure allowed more people to participate in tourism. Many people started to travel abroad, and Japan became a popular destination for foreign tourists. However, during this period, there was also a negative impact on the environment and local culture due to the large influx of tourists (Nobukiyo, 2010).

During the period of 1964–1969, many young and middle-aged Japanese migrated to urban areas to participate in the rapidly growing economy. This caused depopulation in rural areas and overcrowding in urban areas, particularly Tokyo. In 1964, the Japanese government began allowing tourism abroad due to the fixed foreign-exchange rate, allowing the public to exchange no more than US$500. That same year, Japan hosted the Olympic Games, becoming the first Asian country to do so. The development of high-speed transportation, particularly the Shinkansen bullet train, allowed for easier travel between Tokyo and Osaka. Additionally, the first commercial jet flight over Japan in 1969 made air travel more affordable. The middle class, who had limited leisure time and no cars for extensive travel, were likely the target of new parks and health centers built near suburban areas (Nobukiyo, 2010). This trend continued into the next period.

The period of 1970–1980s marked the beginning of mass tourism in Japan, driven by stable economic growth. The World Expo held in Osaka in 1970 attracted a large number of tourists from foreign countries and all over Japan. The event was accompanied by the organization of numerous package tours, which further boosted tourism in Japan. After the Expo ended, National Railways launched the "Discover Japan" campaign, promoting traditional images such as Shinto shrines, Buddhist temples and traditional dances of rural towns and villages, which contributed significantly to domestic tourism in rural areas. Additionally, the construction of ski resorts and golf courses in Honshu and Hokkaido proved popular among young people and wealthy or business people, respectively. However, the oil crises of 1973 and 1978 forced Japanese society to rethink the impact of mass tourism on the natural environment (Nobukiyo, 2010).

From the 1980s to 2006, a new type of tourism emerged in Japan known as "alternative" tourism, which was the result of the diversification of Japanese people's lifestyles and increasing demand for tourism activities in resorts and on cruise ships, among other things. Despite the bubble economy that Japan experienced in the late 1980s and early 1990s, tourism continued to grow gradually. In 1990, the number of outbound visitors passed ten million for the first time, and inbound visitors exceeded three million. Although there were negative events such as the September 11 attacks in the United States in 2001, the SARS outbreak in 2004, and the severe economic depression in 2008–2009, inbound tourism has continued to grow and reached approximately 31 million international visitors in 2018. Meanwhile, outbound tourism growth has been unstable since the 2000s, as shown in Figure 1.2

As mentioned previously, the diversification of Japanese people's lifestyles and demand for different types of tourism led to the rise of alternative tourism in Japan. Various types of alternative tourism, such as ecotourism, rural tourism, green tourism and blue tourism, emerged as a result, with tourism industries, tourists, local governments and communities all creating new destinations. However, each tourism stakeholder, such as travel agencies and tour operators, approached and exploited new types of tourism in different ways. For example, tourism industries focused on maximizing benefits while minimizing costs by using new labels for tourism. Travel agents sold eco-tours and mountain visits as ecotourism (Nobukiyo, 2010). One example of this is Yakushima Island, the first Japanese designated Natural World Heritage Site in 1993, which has experienced an uncontrolled influx of tourists, leading to negative impacts on the natural environment (Forbes, 2012). Therefore, achieving any form of alternative tourism is challenging. It is necessary not only to manage tourism destinations by controlling the number of tourists but also to educate tourists on ethical behavior (Weeden, 2013).

Since 2006, Japan has been working toward becoming a tourism nation, with the establishment of the Japan Tourism Agency (JTA), an extra-ministerial bureau founded by the Ministry of Land, Infrastructure, Transport and Tourism (MLIT) in 2008. The JTA's mission is to stimulate local economies and promote international mutual understanding, in line with the "Tourism Nation Promotion Act" passed in December 2006. According to a JTA report from 2009, the "Tourism Nation Promotion Plan" had five key targets: (1) increasing the number of foreign tourists visiting Japan, (2) increasing the number of Japanese travelers visiting overseas, (3) increasing the number of international meetings held in Japan, (4) increasing

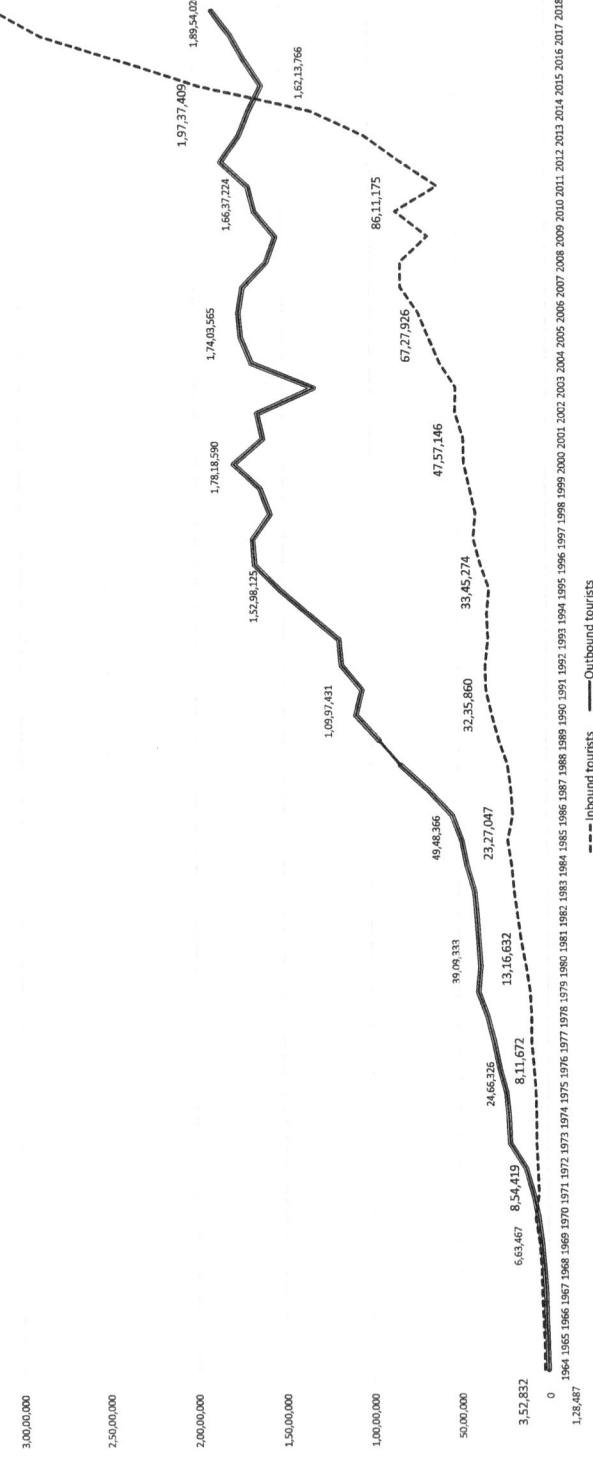

Figure 1.2 Changes in inbound and outbound tourism in Japan (1964–2018).

Source: Japan National Tourism Organization (2019)

the number of nights Japanese tourists spend on accommodation during domestic sightseeing tours, and (5) increasing expenditure on sightseeing tours in Japan. The first target has been successful, as the number of international tourists visiting Japan has been steadily increasing (see Figure 1.2).

This section provides a brief overview of the history of tourism development in Japan. It highlights that tourism has been present in Japan since the 19th century and initially served the purpose of pilgrimages and health-related reasons, although these purposes still persist today. Over time, tourism in Japan has shifted its focus toward economic growth, particularly in recent years. Additionally, this section briefly touches on the tourism policies that have contributed to Japan's tourism development but will be discussed in greater detail in Section 1.3.

1.2 Poverty in Japan

As previously discussed, Japan is also facing issues of poverty, particularly in relation to relative poverty or income inequality. Over the past few decades, Japan has seen a significant shift in its poverty and income distribution landscape. In the 1970s, the country was known as a "society of 100 million middle-class people". However, according to the Ministry of Health, Labour, and Welfare's Survey on the Redistribution of Income, the Gini coefficient, which measures income inequality on a scale of 0 to 1, increased by 11% between the mid-1980s and 2000, indicating a rise in inequality (Jones, 2007).

On the other hand, a survey conducted by the Cabinet Office in 2012 found that 92.3% of Japanese residents still considered themselves members of the middle class in terms of living standards. However, this number has recently decreased significantly, with only 56.6% of respondents identifying as middle class in the same survey in 2016 (Cabinet Office, 2016). This decline likely reflects the long-term effects of economic and social changes that occurred after the collapse of the bubble economy in the early 1990s, as well as changes in the employment system, such as an increase in nonpermanent employment.

Statistics from the OECD indicate that while poverty rates and income inequality in Japan have decreased since the 2000s, they were higher than the OECD average from the mid-1980s to the mid-2000s (OECD, 2008). The data also indicate that Japan's income inequality and poverty rates are among the highest of all OECD member countries, with a top 16 ranking for income inequality (i.e., Gini coefficient 2015: 0.33) and a poverty rate of 16.1% in 2012. These figures suggest that poverty is relatively high among young and elderly people in Japan, and that one in six Japanese people are living in relative poverty (see Figure 1.3).

Poverty rates and income inequality in Japan are not uniform across the country's prefectures. Income inequality in Japan is influenced not only by differences between the working-age population (such as the wage distribution between regular and nonregular workers) but also by regional differences (such as the Kanto region compared to other regions) and urban–rural differences (such as the high concentration of people in metropolitan areas). This results in different costs of living across these areas. Several factors contribute to income inequality, as described by Tachibanaki (2006). These factors include changes in wage payment methods

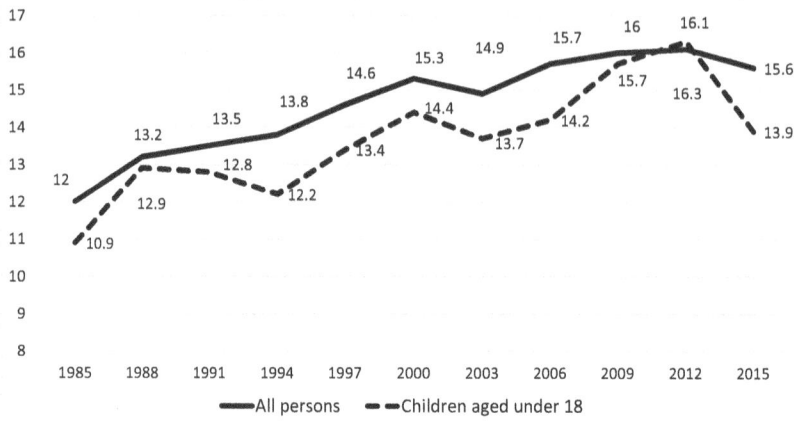

Figure 1.3 Changes in relative poverty rate in Japan (1985–2015).

Source: Ministry of Health, Labor and Welfare (2017)

(from the seniority payment principle to the performance-based payment principle), an aging population that creates income gaps between young and old people
and changes in family structure that widen household incomes (such as a decrease
in the number of family members and an increase in the number of single-person
households).

The factors mentioned earlier suggest that measuring poverty by a single standard may be insufficient or inadequate. Therefore, it is necessary to consider income
inequality not only in terms of actual distribution but also in terms of how people
perceive or feel about inequality (Mizuho Research Institute, 2015).

1.3   Tourism policies in Japan

Together with the growth of tourism, tourism policy in Japan also has a long history. This section briefly chronicles the development of Japanese tourism policies
after the Second World War and then examines the important frameworks of tourism policy in modern Japan.

1.3.1   The development of tourism policies in Japan

The development of tourism policies in Japan can be divided into four main periods, each with its own specific characteristics of tourism policies.

In the Early Period after the Second World War until 1964, the Japanese government's perception of tourism development underwent changes. Travel into
and out of the country was strictly forbidden by the Occupation forces until
December 1947, when American tourists travelling to the Far East on board the
liner *President Monroe* were allowed to disembark at Yokohama for 24 hours

(Funck & Cooper, 2013). During the Occupation, only several small groups of American tourists were permitted to visit Japan. However, after Japanese manufacturers began to export successfully overseas, the Japanese government recognized inbound tourism as a major foreign-exchange earner. This led to the emergence of a number of travel agencies such as Nihon Tourist in 1948 and the promulgation of the Travel Intermediary Law in 1952.

One of the most notable tourism policies during this period was the promulgation of the Basic Law on Tourism in 1963, a year before the Tokyo Olympic Games in 1964. This law paved the way for the liberalization of economic policies on travel and emphasized the importance of projecting a new image of Japan worldwide. As a result of Japan's emerging political position and its rapid economic growth as an industrialized nation, the previous restrictive regulations were abolished, allowing Japanese people to travel abroad freely (Funck & Cooper, 2013).

The period from 1964 to the 1990s witnessed strong growth in both inbound and outbound tourism in Japan. To maintain the nation's high economic growth and political status, the government decided to open up the outbound travel market, allowing all Japanese people to travel overseas freely from April 1, 1964. While the government also attempted to promote inbound tourism, outbound tourism was more attractive to Japanese travellers due to the strength of Japan's currency, the yen, at that time (Funck & Cooper, 2013). One of the key organizations established during this period was the Japan National Tourist Organization (JNTO), which was created by law in April 1964. This nonprofit statutory organization is under the supervision of the MLIT and is designed to promote inbound travel to Japan, as well as deepen travellers' understanding of Japanese history, culture, traditions, customs and people (JNTO, 2006).

Third, from 1997 onwards, there was a focus on inbound tourism policies in Japan, and the government played an important role in attempting to increase the number of international visitors to Japan. This was not only for economic reasons but also to enhance mutual understanding between people from different countries and cultures. Various campaigns, plans and programs were formulated by different tourism stakeholders during this period to promote inbound tourism, for instance, the "Welcome Plan 21" in 1997, the "Visit Japan" Campaign by JNTO in 2001, the "East Asia Sphere for Tourism Plan" in 2001, the "China-Japan Mutual Visit Year" in 2002, "Bilateral Consultation on expansion of tourism in the Early 21st Century" and the "Yokoso! Japan" Campaign in 2003, among others (OECD, 2002).

Fourth, the post-2007 period marked a significant increase in inbound tourism in Japan as a result of the policies mentioned earlier. By 2007, international tourist arrivals had reached about eight million. However, this number slightly declined in the aftermath of the global financial crisis that occurred in 2008. To counter the resulting decline in both inbound and outbound tourism, the Japan Association of Travel Agents (JATA) launched the Visit World Campaign to revitalize the Japanese market. The campaign aimed to increase the number of Japanese outbound travellers to 20 million by 2010 and strengthen the business relationship between Japanese tour operators and travel agencies and their overseas counterparts. It also sought to promote nine major destinations (South Korea, Hong Kong, Thailand,

Taiwan, Guam, Australia, the US mainland and Hawaii and France) as well as three new destinations (China, Vietnam and Macao) for travel agents and the government to focus on (JATA, 2008).

1.3.2   The framework of tourism policy in modern Japan

The organization of national tourism administration in Japan has undergone several changes over the years. In 2001, the Ministry of Land, Infrastructure and Transport (MLIT) was established, and tourism promotion and development were included in this administration. This was done because tourism is closely related to transport policy, including air, land and maritime transport, as well as the provision of infrastructure, regional development policy and the goal of achieving a higher quality of life.

Within this structure, a Department of Tourism was set up under the Policy Bureau of MLIT, which is responsible for tourism policy and policy coordination on behalf of the Japanese government. Under the direction of the Director-General of Tourism, three divisions were organized: the Planning Division, responsible for the total coordination of tourism policy, research and planning, promotion of inbound tourism and international affairs; the Regional Development Division, responsible for regional development by tourism promotion, provision of tourism-related facilities, sustainable development of tourism and registered hotels and ryokan; and the Travel Promotion Division, responsible for the supervision of travel agents, development of tourism industries, promotion of tourism demand for Japanese citizens and consumer protection (OECD, 2002).

In terms of the major goals of tourism policies, as noted previously from 1997 onwards, the strategic framework has been primarily focused on promoting foreign visitors. However, there was also a framework in place to encourage travel demand within Japan. For example, the Ministry of Land, Infrastructure and Transport launched a new vision for regional development in 2001 called "Tourism Based Community Development". Traditionally, tourism development placed a priority on increasing visitors and contributing to the industry, while paying less attention to the impacts of tourism on local people and the environment. In this new vision, three key factors were identified for sustainable tourism development: (1) resources, (2) the living environment, and (3) visitor satisfaction. Sustainable development of the entire community can be achieved while maintaining a balance of these three key factors through integrated community development, in which the community plays a central role in tourism promotion (OECD, 2002).

In the early 21st century, the Japanese government recognized the importance of promoting tourism development in local and regional areas. As a result, programs such as the "Tourism Exchange Space Model Projects" and the "Tourism Plus 1 Strategy" were developed in order to assist local communities in leveraging their unique characteristics to attract tourists. Another project, called the "Tourism Renaissance", was launched in 2005 with the aim of supporting private sector tourism promotion efforts, conducting basic research, developing local brands, fostering human resources and facilitating information transmission. Additionally, according to the Ministry of Land, Infrastructure, Transport and Tourism's (MLIT's) 2016 report on the "New tourism strategy to invigorate the Japanese economy", regional revitalization was identified as a crucial aspect of this strategy.

In 2007, the Japanese government enacted the Tourism Nation Promotion Basic Law, which identified tourism as an essential foundation of Japan's national policy in the 21st century, replacing the Basic Law of 1963. As a result, the Cabinet Office proclaimed the Tourism Nation Promotion Basic Plan on June 29th, which sets various targets, including increasing the number of international visitors to Japan to ten million, increasing the number of Japanese tourists traveling abroad to 20 million, increasing the value of tourism consumption to JP¥30 trillion, increasing the number of overnight stays per person for domestic travel to four nights per year and increasing the number of international conferences held in Japan by at least 50% (MLIT, 2012).

To achieve these targets and promote Japan as a tourism nation, the government committed itself to these efforts. Toward this end, the Ministry of Land, Infrastructure and Transport was recently reorganized and expanded to become the Ministry of Land, Infrastructure, and Transport and Tourism, and the Japan Tourism Agency (JTA) was established in 2008 (OECD, 2014; Funck & Cooper, 2013).

The section primarily highlights the presence of numerous tourism policies implemented in Japan. These policies not only serve economic growth but also aim to address the challenge of imbalanced regional development, such as disparities between central regions like the Kanto region and less-favored regions like the Chugoku or Shikoku regions. However, it is crucial to note that these policies are not directly linked to poverty alleviation or the role of tourism in promoting pro-poor tourism. Moreover, this indicates a lack of attention toward PPT, particularly in terms of addressing the unequal distribution of benefits within the tourism sector, in the case of Japan.

Reference list

Cabinet Office. (2016). *Public opinion survey on the life of the people*. Retrieved from www.gov-online.go.jp/eng/pdf/summaryk16.pdf

Forbes, G. (2012). Yakushima: Balancing long-term environmental sustainability and economic opportunity. *Kagoshima Immaculate Heart College Research Bulletin, 42*, 35–49.

Funck, C., & Cooper, M. (2013). *Japanese tourism: Spaces, places and structures*. New York, NY: Berghahn.

Japan Association of Travel Agents (JATA). (2008). *Visit world campaign. 2008 promotional activity report and plan for 2009*. Retrieved May 2019, from www.jata-net.or.jp/vwc/pdf/2008report_2009plan.pdf

Japan National Tourism Organization (JNTO). (2006). *JNTO—What we do*. Retrieved from www.jnto.go.jp/eng/about/pdf/about_JNTO_20060925.pdf

Japan National Tourism Organization (JNTO). (2019). *Japan tourism statistics*. Retrieved May 2019, from https://statistics.jnto.go.jp/en/graph/

Jones, R. (2007). *Income inequality, poverty and social spending in Japan* (Economics Department Working Papers, No. 556). Paris: OECD.

Ministry of Health, Labor and Welfare. (2017). *Comprehensive survey of living conditions 2015*. Retrieved May 2019, from www.mhlw.go.jp/english/database/db-hss/cslc-tables.html

Ministry of Land, Infrastructure, Transport and Tourism (MLIT). (2012). *Tourism nation promotion basic plan*. Tokyo: Cabinet Office.

Mizuho Research Institute. (2015). Japan's inequality today and policy issues. *Mizuho Economic Outlook and Analysis*. Retrieved from www.mizuho-ri.co.jp/publication/research/pdf/eo/MEA151007.pdf

National Parks of Japan. (2019). Retrieved from www.japan.travel/national-parks/parks/yakushima/

Nobukiyo, E. (2010). A brief review of tourism in Japan after World War II. *Journal of Ritsumeikan Social Sciences and Humanities, 2*(90), 141–153.

Organization for Economic Co-operation and Development (OECD). (2002). *National tourism policy review of Japan*. Paris: OECD.

Organization for Economic Co-operation and Development (OECD). (2008). *Growing unequal? Income distribution and poverty in OECD countries*. Paris: OECD. Retrieved from www.oecd.org/japan/41527303.pdf

Organization for Economic Co-operation and Development (OECD). (2014). *"Japan", in OECD tourism trends and policies 2014*. Retrieved April 2019, from http://dx.doi.org/10.1787/tour-2014-25-en

Tachibanaki, T. (2006). Inequality and poverty in Japan. *The Japanese Economic Review, 57*(1), 1–27.

Weeden, C. (2013). *Responsible and ethical tourist behaviour*. London, UK: Routledge.

2 Tourism development and poverty in Vietnam

2.1 Tourism development: an overview

2.1.1 History of tourism development

In order to gain a comprehensive understanding of Vietnam's tourism sector, it is crucial to examine its history and development, as noted by scholars such as Huynh (2011) and Tran (2005). Vietnam's tourism industry has undergone more than 50 years of evolution, which can be divided into three distinct periods (Figure 2.1 shows the location of Cu Lao Cham, Vietnam). During the 1960s to 1975, tourism was primarily developed for political purposes, as noted by Brennan and Nguyen (2000) and Tran (2005). From 1976 to 1990, the economic potential of tourism was recognized and explored, according to Cooper (2000) and Tran (2005). Since the 1990s, tourism has been utilized as a means of economic growth and poverty reduction, as highlighted by the government (GOV, 2005a).

From 1960 to 1975, Vietnam was divided into two regions, North and South, during the American War, also known as the Vietnam War in the West. According to the Vietnam National Administration of Tourism (VNAT) in 2005, developing tourism during this time was challenging since it primarily served political objectives. The majority of foreign tourists were political delegates who were invited by the Vietnamese government, and it was uncommon to see leisure or business tourists during this period. Consequently, the total number of international arrivals was extremely limited (refer to Figure 2.2). In 1960, the Vietnam Tourist Company, the country's first tourist company, was established in the North under the management of the Ministry of Foreign Affairs. Later, it came under the control of the Ministry of Public Security, as noted by Tran (2005) and VNAT (2005). The available evidence indicates that the state played a crucial role in tourism management, and economic benefits were not a top priority, as observed by Cooper (2000). Despite some existing tourist sites such as Hanoi, Hai Phong, Hoa Binh and Tam Dao, it can be argued that tourism was not yet an economic sector or an economic activity during this period (Truong, 2014).

Vietnam faced various obstacles in the period from 1976 to 1990 when trying to develop its tourism industry. The aftermath of the American War had resulted in substantial destruction (Mok & Lam, 2000). Nevertheless, tourist destinations in

DOI: 10.4324/9781003496748-4

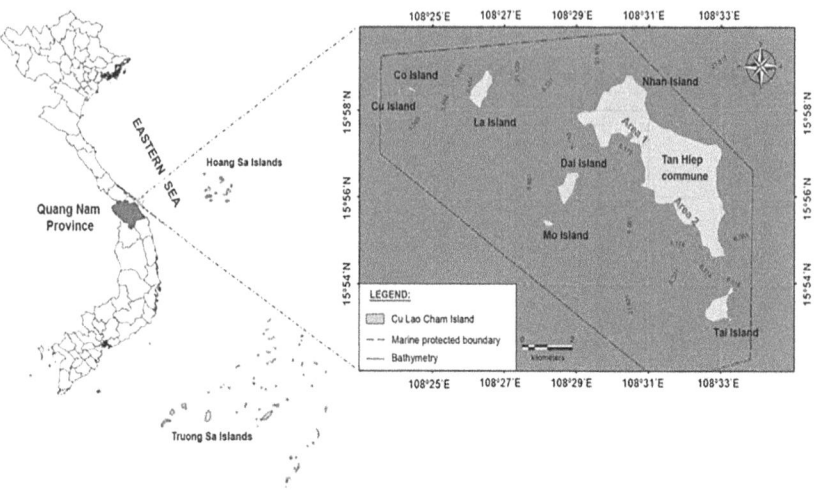

Figure 2.1 Location of Cu Lao Cham in Vietnam.
Source: Tin et al. (2020)

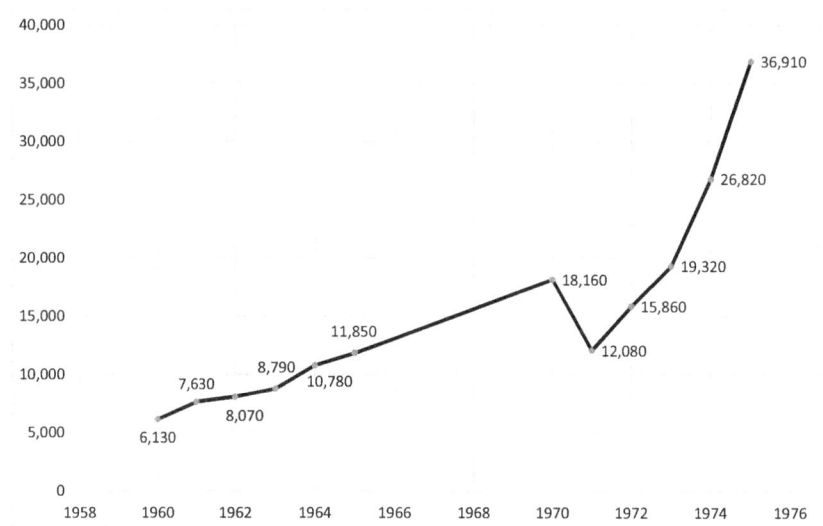

Figure 2.2 International arrivals to the North region of Vietnam, 1960–1975 (unit: person).
Source: (Tran, 2005)

the southern region, such as Ho Chi Minh City, Hue, Da Nang, Vung Tau and Can Tho, gradually expanded during this time (VNAT, 2005). Local people's committees established and managed several state-owned tourist companies in these areas. However, the number of international tourists remained limited during this period, as shown in Figure 2.3. Most of the visitors were from the former Soviet Union

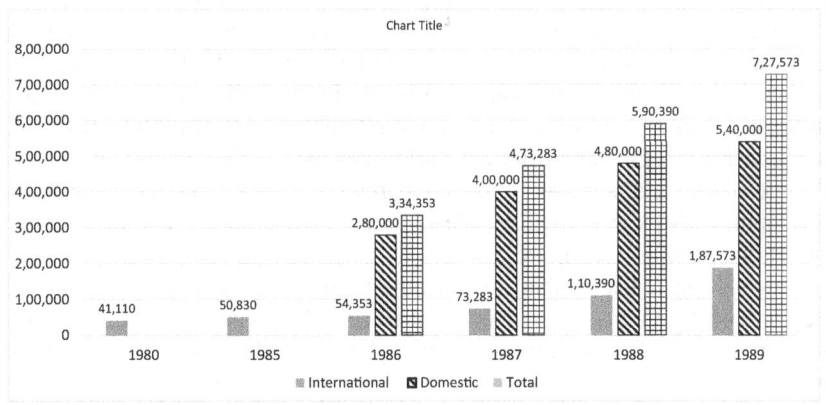

Figure 2.3 Tourist arrivals to Vietnam, 1980–1989 (unit: person).

Sources: (Tran, 2005)

Table 2.1 International arrivals 1990–1999 (unit: thousand)

Year	1990	1991	1992	1993	1994	1995	1996	1997	1998	1999
Total	250	300	440	670	1,018	1,351	1,067	1,715	1,520	1,782
By purposes										
Leisure	–	–	–	–	–	610	661	691	599	838
Trade & investment	–	–	–	–	–	308	364	403	292	266
Visit friends & relatives	–	–	–	–	–	–	273	371	301	337
Others	–	–	–	–	–	432	306	249	328	341
By means of transport										
Airways	–	–	–	–	–	1,206	939	1,033	874	1,022
Waterways	–	–	–	–	–	21	161	131	157	188
Land	–	–	–	–	–	122	505	550	489	572

Source: VNAT (2018)

(Cooper, 2000; VNAT, 2005). After the reunification in 1975, Vietnam aimed to develop tourism as a means to foster patriotism, facilitate reconciliation between the North and South and present itself as a peaceful nation to the global community (VNAT, 2005). The introduction of the Renewal Policy (*Doi Moi*) in 1986 marked a significant turning point in Vietnam's development by transitioning the country from a command economy to a market economy. Although tourism started attracting more foreign visitors, it was still considered a secondary sector and did not experience substantial growth until the 1990s (Tran, 2005).

According to Brennan and Nguyen (2000), significant changes in tourism development occurred from 1991 to present with the issuance of laws promoting private and foreign investment in Vietnam. The removal of barriers to private investments and encouragement of foreign investments led to a rapid increase in foreign tourists seeking business and investment opportunities (see Table 2.1) (Hobson, Heung, & Chon, 1994; Mok & Lam, 2000). This shift in perception by the government from political to economic, as noted by Cooper (2000), was reflected in the declaration of tourism as a strategic component for socioeconomic development, industrialization and modernization in Decree No.46 of 1994 (VNAT, 2005). More policies and strategies relating to tourism were issued in the 2000s, such as the Tourism Ordinance issued in 1999 and the Law on Tourism issued in 2005, providing a comprehensive legal framework for tourism businesses. The stability of Vietnam's socioeconomic and political system also provided a good foundation for tourism development (Mok & Lam, 2000), leading to a gradual increase in the number of tourists since 2000 (see Figure 2.4).

This section indicates that, historically, tourism has been utilized to achieve the goals of the Vietnamese government since 1960. In each period, tourism was developed with different purposes, such as for political goals from 1960 to 1975, as an economic sector from 1976 to 1990, and for economic growth and poverty alleviation since 1991 to the present day. The background of tourism development in Vietnamese nature conservation areas will be presented next.

2.1.2   *Tourism development in nature conservation areas*

Vietnam has four main categories of protected areas (PAs) that often overlap: national parks, nature reserves, species/habitat reserves and protected landscapes (which include cultural sites). Marine protected areas (MPAs) in Vietnam are categorized as "other protected areas" as they have only been proposed in recent years (i.e., the first MPA was established in 2001) (Clarke, 1999; ICEM, 2014). As of 2017, Vietnam has a total of 167 protected areas and 18 MPAs (Ha, 2018; Ministry of Agriculture and Rural Development (MARD), 2017).

Although the first protected area, Cuc Phuong National Park, was established in 1962 (ICEM, 2003), tourism development in Vietnamese protected areas has only recently been recognized. The legal framework for tourism development in protected areas is contained in Article 16 of Decision 08/2001/QD-TTg. This states that "protected area management boards can organize, lease out, or contract the provision of ecotourism services and facilities to organizations, households, and individuals, in compliance with existing financial management regulations and subject to a majority of earnings being reinvested in managing, protecting, and developing the protected area" (ICEM, 2003, p. 42).

However, several issues still exist regarding tourism development in Vietnam's PAs (ICEM, 2003; GIZ, 2011; Pham, 2017). First, there is a lack of an institutional framework. It is unclear which organizations (e.g., management boards of PAs, MARD, VNAT, or local governments) should be responsible for tourism development in PAs due to its complicated management system. The involvement of private sectors regarding tourism in PAs is also not clearly stated. Private-sector tourism needs to be governed by a regulatory framework to avoid some negative impacts on local communities. Likewise, tourism benefits should be shared equitably with local people (Ly & Xiao, 2016; Pham, 2017; GIZ, 2011).

Second, there is a lack of reliability of tourism benefits to local communities. Internationally, tourism has a bad image for "economic leakage", that is, much of

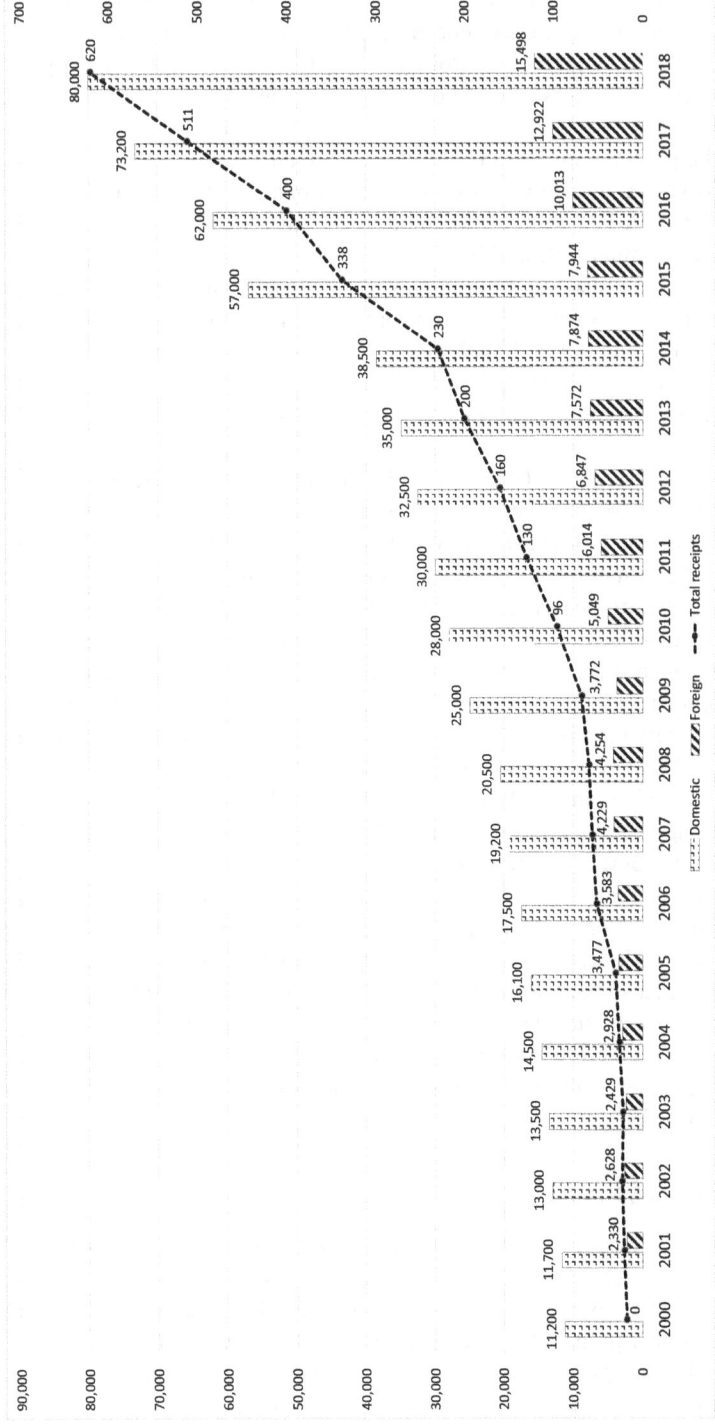

Figure 2.4 Vietnam's tourism arrivals (thousand) receipts (trillion VND) 2000–2018.

Source: VNAT (2018); 620 trillion VND is equivalent to about US$ 31 billion

the profits are not retained at the tourist site (ICEM, 2003; Sandbrook, 2010). Very often, tourist operators use experienced staff from outside the area instead of training local people. Transport to and from the tourist site is also undertaken by outside operators. Retaining tourism revenues locally is essential for local communities to benefit from it.

Third, the potential attractions of many of Vietnam's PAs are limited. Vietnam does not possess abundant wildlife that international tourists will pay to see, and tourist facilities in PAs (e.g., accommodations, information centers, guides, and trails) are very limited. About 70% of the leisure and tourist destinations of Vietnam are located in coastal areas, attracting about 80–90% of tourists per year (Do & Phi, 2022). There are very few coastal protected areas that develop tourism services. Vietnam's PAs attract largely domestic tourists. In recent years, many tourists to PAs have been students on school outings, suggesting a great potential for environmental education tourism for young people (Do & Phi, 2022; Lanh & Truong, 2018).

According to MARD (2017), currently, around 61 out of Vietnam's 167 protected areas have organized ecotourism activities, including 25 out of 34 national parks and 36 out of 133 nature reserves. However, in reality, tourism development in PAs lacks fundamental master plans, leading to negative impacts on nature, local cultures, and the effectiveness of tourism businesses in the PAs. For instance, out of the 61 Vietnam's PAs that have operated ecotourism, only one PA had a tourism investment project approved by the government, and there were only tourism development projects/proposals in five PAs (MARD, 2017). This suggests that while the conservation of biodiversity values is the main focus of Vietnam's PAs based on the Vietnam Biodiversity Law (GOV, 2008), tourism is not considered a primary strategy in many of these areas. Moreover, the institutional framework for tourism management in the Vietnamese PAs faces various issues, which limits the development of tourism in these areas.

One of the challenges of tourism in Vietnam's protected areas is ensuring that local communities benefit from it. Sustainable tourism in these areas can be more advantageous for local development if the people who live in or around the protected areas can also benefit from it. This can be achieved by creating employment opportunities in tourism, such as through homestays, selling handicrafts, providing tourist services and so on. If local communities only see the costs of protected areas and tourism, rather than the benefits, they may not support them (ICEM, 2003; Bennett & Dearden, 2014).

2.2   Poverty in Vietnam

As previously discussed, Vietnam suffered severe damage after the American War. The American government imposed economic sanctions on Vietnam, making the country's economy even more challenging in the following years (Dang, 2009). Vietnam was also politically isolated from the West, and as a result, it was one of the world's poorest countries until the mid-1980s. The country faced various obstacles to economic growth, including negative per capita income, widespread famine, and low investment. It heavily relied on financial aid from the former Soviet Union (Dollar & Litvack, 1998).

Although the Vietnamese government attempted to develop the economy by implementing collectivized agriculture and subsidizing state-owned industries, these efforts yielded limited results (Dollar & Litvack, 1998). Therefore, in 1986, the government adopted the *Doi Moi*, which transformed the centrally planned economy into a market-oriented one.

The Vietnamese government initially implemented the *Doi Moi* in agriculture, which is the largest sector of the economy. In 1988, the government abolished agricultural collectives, and as a result, farming land was distributed to households. In 1993, the government issued a new land law that granted farmers the right to use their allocated land and renew these rights after 20 years (Dollar & Litvack, 1998).

Other economic sectors were also reformed effectively. For instance, state-controlled commodity prices were abolished, and products could be sold at market prices, which encouraged private production of goods and services to be involved in the market. By the end of 1989, there was rapid growth in agriculture, services and construction—three areas where the private sector responded quickly to the state incentives (Dollar & Litvack, 1998). In the 1990s, structural reforms were also implemented in other sectors such as finance and banking.

The *Doi Moi* helped to reduce the number of people living in poverty in Vietnam. The country's poverty rate decreased from 75% in 1990 to 55% in 1993 (Dollar & Litvack, 1998). Vietnam was recognized for its success in halving the poverty rate between 1990 and 2000 (UNDP, 2009). Currently, Vietnam is ranked 116th in terms of the Human Development Index, which is at the medium level. The Human Development Index comprises three main dimensions: a long and healthy life, being knowledgeable and having a decent standard of living (UNDP, 2018).

Despite Vietnam's significant achievements in economic growth and poverty reduction, it remains a relatively poor country (GOV, 2003; Huxford, 2010). Poverty is defined as "a situation in which a proportion of the population does not have the means to satisfy basic human needs that have been recognized by society, depending on the level of economic and social development and local customs and practices" (GOV, 2003, p. 17). Poverty is measured using poverty lines that are developed by the GOV for both urban and rural areas. These lines are adjusted over time depending on socioeconomic changes. For instance, the GOV's poverty lines for the period of 2006–2010 were VND400,000/person/month (US$20) in rural areas and VND500,000/person/month (US$25) in urban areas, while those for 2011–2015 were VND480,000/person/month (US$24) and VND600,000/person/month (US$30), respectively. The poverty lines for 2016–2020 are VND700,000 (US$35.0) and VND900,000 (US$45.0) per person per month in rural and urban areas, respectively (GSOV, 2010, 2011, 2017). The GSOV conducts a household living standards survey every two years, and the poverty rates are identified by residence and region, as shown in Table 2.2.

Table 2.2 illustrates that poverty rates in Vietnam vary according to the region and the residence. The highest poverty rates are found in rural and mountainous areas, where ethnic minorities make up the majority of the poor, including the

Table 2.2 Vietnam's poverty rate by residence and by region 2006–2016 (unit: %)

Region Year	2006	2008	2010	2011	2014	2015	2016
Whole country	**15.5**	**13.4**	**14.2**	**12.6**	**8.4**	**7.0**	**5.8**
By residence							
Urban	7.7	6.7	6.9	5.1	3.0	2.5	2.0
Rural	18.0	16.1	17.4	15.9	10.8	9.2	7.5
By region							
Red River Delta	10.0	8.6	8.3	7.1	4.0	3.2	2.4
Northern midlands and mountain areas	27.5	25.1	29.4	26.7	18.4	16.0	13.8
North Central and Central coastal areas	22.2	19.2	20.4	18.5	11.8	9.8	8.0
Central Highlands	24.0	21.0	22.2	20.3	13.8	11.3	9.1
South East	3.1	2.5	2.3	1.7	1.0	0.7	0.6
Mekong River Delta	13.0	11.4	12.6	11.6	7.9	6.5	5.2

Source: GSOV (2011, 2017)

Northern midlands and mountainous areas, the North Central and Central coastal areas and the Central Highlands. These regions are geographically isolated and often have limited access to information, markets, capital and infrastructure (GOV, 2003; WB, 2018). It is worth noting that Vietnam has 54 ethnic groups, with the Kinh people being the largest. The GOV promotes policies of equality, solidarity and mutual assistance among all ethnic groups (ADB, 2002). Additionally, there is a strong correlation between protected areas and poverty. The reasons for poverty in and around Vietnam's protected areas are also related to the same factors of remoteness, mountainous terrain and isolation, which limit access to markets and land (ICEM, 2003; WB, 2018).

In addition, Table 2.2 suggests that poverty rates are low in the Red River Delta and the Mekong River Delta. These regions are home to the capital city of Hanoi and Vietnam's largest city of Ho Chi Minh, respectively, and their residents often have more opportunities for economic involvement and growth.

2.3 Tourism policies and poverty alleviation

This section examines the changes in Vietnam's tourism policies from 1991 to the present day, with a particular focus on poverty alleviation. The periods of 1960–1975 and 1976–1990 have been excluded since tourism was not considered a formal sector during the first stage, and thus, no significant policies were implemented. During the second stage, although a number of policies were introduced, they did not reflect tourism policies that aimed to alleviate poverty.

As noted earlier, the Vietnamese government has considered tourism as an important means of poverty alleviation since the 1990s. To promote tourism development, the government has formulated several strategies and policies, as summarized in Table 2.3. However, this section focuses only on those policies, strategies and plans that have a poverty component. These include the "Master Plan for Tourism Development 1995–2010", the "Tourism Ordinance", the "Comprehensive

Table 2.3 Tourism development policies, strategies and plans

Publication	Year of approval	Main objective	Poverty component
Master Plan for Tourism Development 1995–2010	1994	By 2000: 3.5–3.8 million foreign tourists; 11 million domestic tourists; turnover US$2.6 billion. By 2010: 9 million foreign tourists; 25 million domestic tourists; turnover US$11.8 billion	Sustainable tourism emphasized by protecting tourism resources and creating economic opportunities for disadvantaged regions
Tourism Ordinance	1999	Tourism considered as an important industry that improves intellectual standards, creates employment, and contributes to socioeconomic development	Tourism encouraged in socially and economically backward regions
National Action Plan for Tourism Development 2002–2005	1999	Ensure that tourism would be a key industry and turn Vietnam into an advanced tourist destination by 2005. Receive 3–3.5 million foreign tourists and 15–16 million domestic tourists by 2005	Sustainable tourism mentioned. Poverty alleviation neglected
Socioeconomic Development Strategy 2001–2010	2001	Lift Vietnam out of the state of underdevelopment, improve local living standards, and lay the foundation for Vietnam to become an industrialized country by 2020	Efforts encouraged to develop tourism into a spearhead industry. Poverty alleviation ignored
NSTD 2001–2010	2002	Develop tourism into a spearhead industry and turn Vietnam into an important tourist destination in Asia	Effective use of tourism resources encouraged, but poverty alleviation not mentioned
CPRGS	2003	Lift Vietnam out of the state of underdevelopment and lay the foundation for it to become an industrialized country by 2020	Rapid and sustainable economic growth emphasized as the best way out of poverty

(Continued)

Table 2.3 (Continued)

Publication	Year of approval	Main objective	Poverty component
Law on Tourism	2005	Issue policies to ensure that tourism could become a spearhead industry	Tourism encouraged in remote regions for hunger elimination and poverty reduction
National Action Plan for Tourism Development 2006–2010	2006	Turn Vietnam into an advanced tourist destination in Asia by 2010. Foreign tourists increase about 10–20% annually. Domestic tourists increase about 15–20% annually	Poverty alleviation not mentioned
National Action Plan for Tourism Development 2007–2012	2007	Achieve targets set by NSTD 2001–2010: develop tourism into a spearhead industry; receive 5.5–6 million foreign tourists and 25 million domestic tourists	
NSTD to 2020 (vision 2030)	2011	Affirm tourism as the spearhead industry and use total tourism receipts as the primary indicator of tourism development	Many action plans made to promote tourism growth. Only one plan proposed for PPT development

Source: Truong (2014)

Poverty Reduction and Growth Strategy" (CPRGS), the "Law on Tourism", and the "National Strategy for Tourism Development (NSTD) to 2020 (vision 2030)".

Along with *Doi Moi*, tourism was seen as a crucial means for Vietnam to integrate into the global economy (Brennan & Nguyen, 2000). The objectives of the aforementioned policies suggested that tourism was considered the best solution for gaining economic benefits and addressing economic issues facing Vietnam (Cooper, 2000). In fact, Vietnam required a significant amount of foreign currency for industrialization and modernization (Cooper, 2000; Nguyen, 2002). Internationally, tourism serves as a significant foreign-exchange earner in many countries, and Vietnam developed its tourism sector to meet those needs.

Until the late 1990s, the *Tourism Ordinance* was promulgated, stating that "tourism is an important, integrated industry . . . tourism development aims to respond to the demands of tourism, leisure, and recreation of the Vietnamese people and international visitors; to contribute to improvements in intellectual standards, employment generation, and socioeconomic development" (translated from Article 1, GOV, 1999a). The Tourism Ordinance also stated that "all tourist activities having negative impacts on the environment, cultural values and traditional customs, national independence and sovereignty, defense and security are forbidden" (translated from Article 8, GOV, 1999a). This suggests that, on the one hand, the GOV considered tourism as an economic sector to promote development. On the other hand, the GOV was afraid of losing political power in the transitional process and was sensitive to the potential negative consequences of tourism. Therefore, tourism was placed at the national administration level. The National Steering Committee for Tourism Development was established in 1999, headed by the Deputy Prime Minister and including the Deputy Ministers of Public Security and National Defense on its panel (GOV, 1999b).

In the 2000s, Vietnam began to connect tourism with poverty alleviation, as noted by Nguyen et al. (2007). Recognizing that tourism could have significant impacts on the wider population, particularly the poor, various development plans and strategies incorporated tourism. For instance, the *CPRGS* stated, "The Government of Vietnam takes poverty reduction as a cutting-through objective in the process of country socioeconomic development. Vietnam has also declared its commitment to implement the Millennium Development Goals and poverty reduction objectives that had been agreed upon in the National Summit in September 2000" (GOV, 2003, p. iii). The GOV believed that achieving high and sustainable economic growth was essential to narrowing the economic gap between Vietnam and other countries in the world and that high economic growth was the best solution for poverty alleviation (GOV, 2003).

The *CPRGS* also emphasized that poverty alleviation was not solely based on support from the GOV or wealthy entities, but it was also the responsibility of poor people (GOV, 2003). This suggests that the active participation of poor people in helping themselves out of poverty is essential for poverty alleviation to be achievable.

In 2005, the *Law on Tourism* was enacted, which stated that "the State shall create mechanisms and adopt policies to mobilize every resource for increased investment in tourism development to ensure tourism is a national spearhead industry" (Article 6, GOV, 2005a, p. 8). The government also emphasized the importance of developing tourism in "remote and isolated areas and in areas with socio-economic difficulties" to combat hunger and poverty (Article 6, GOV, 2005a, p. 9). This indicates that the government is not only considering tourism as an important industry, but also focused on promoting tourism development in disadvantaged areas (such as rural and mountainous areas) where many poor people reside, as noted previously.

Prior to approving the *NSTD up to 2020 (Vision 2030) in 2011, two National Action Plans for Tourism Development (2006–2010 and 2007–2012)* were developed, but poverty alleviation was not mentioned in these plans. The NSTD to 2020 (Vision 2030) stressed that the main orientation is to make tourism a spearhead sector to contribute to GDP growth and stimulate other industrial sectors simultaneously (GOV, 2013). Importantly, the strategy emphasizes that tourism is developed to generate employment and contribute to poverty alleviation (GOV, 2013, p. 46).

The above discussion shows that although the GOV's tourism policies have considered tourism as an important tool for economic growth since the 1990s, the task of poverty alleviation in the strategies and plans has often been secondary to tourism growth. The GOV pays more attention to targeting the increase of total foreign tourists and tourism receipts. This suggests that the approach to tourism development in Vietnam still emphasizes economic growth rather than poverty alleviation.

Reference list

Asian Development Bank (ADB). (2002). *Indigenous peoples/ethnic minorities and poverty reduction in Vietnam*. Manila: ADB.

Bennett, N. J., & Dearden, P. (2014). Why local people do not support conservation: Community perceptions of marine protected area livelihood impacts, governance and management in Thailand. *Marine Policy, 44,* 107–116.

Brennan, M., & Nguyen, N. B. (2000). *Vietnamese tourism: The challenges ahead*. Retrieved May 2019, from http://ir.lib.oita-u.ac.jp.

Cooper, M. (2000). Tourism in Vietnam: Doi Moi and the realities of tourism in the 1990s. In C. M. Hall & S. Page (Eds.), *Tourism in South and Southeast Asia: Issues and cases* (pp. 167–177). Oxford: Butterworth-Heinemann.

Clarke, J. E. (1999): *Biodiversity and Protected Areas Vietnam*. Regional Environmental Technical Assistance 5771. Poverty Reduction & Environmental Management in Remote Greater Mekong Subregion (GMS). Watersheds Project (Phase I).

Dang, H. L. (2009). *Non-Governmental Organisations (NGOs) and development: An illustration of foreign NGOs in Vietnam* (Master's Thesis, Ohio University, Ohio). Retrieved May 2019, from http://etd.ohiolink.edu.

Do, H., & Phi, G. T. (2022). Marine and island tourism" Stakeholder involvement in policy formulation and implementation. *Vietnam Tourism: Policies and Practice*, 63–84.

Dollar, D., & Litvack, J. (1998). Macroeconomic reform and poverty reduction in Vietnam. In D. Dollar, P. Glewwe, & J. Litvack (Eds.), *Household welfare and Vietnam's transition* (pp. 1–28). Washington, DC: The World Bank.

General Statistics Office of Vietnam (GSOV). (2010). *Statistical yearbook of Vietnam*. Hanoi: Statistical.

General Statistics Office of Vietnam (GSOV). (2011). *Statistical yearbook of Vietnam*. Hanoi: Statistical.

General Statistics Office of Vietnam (GSOV). (2012). *Statistical yearbook of Vietnam*. Hanoi: Statistical.

General Statistics Office of Vietnam (GSOV). (2017). *Statistical yearbook of Vietnam.* Hanoi: Statistical.

German International Cooperation (GIZ). (2011). Co-management/Shared Governance of Natural Resources and Protected Areas in Viet Nam. *Proceedings of the National Workshop on Co-management Concept and Practice in Viet Nam Soc Trang*, 17–19 March 2010.

Government of Vietnam (GOV). (1999a). *Tourism ordinance.* Hanoi: National Political Publishers. (In Vietnamese).

Government of Vietnam (GOV). (1999b). *Decision 23/TTg on the establishment of the national steering committee for tourism development.* Retrieved April 2019, from www.vietnamtourism.gov.vn. (In Vietnamese).

Government of Vietnam (GOV). (2003). *Comprehensive poverty reduction and growth strategy* (CPRGS). Hanoi: Cartography Publishers.

Government of Vietnam (GOV). (2005a). *Law on tourism.* Hanoi: National Political Publishers. (In Vietnamese).

Government of Vietnam (GOV). (2013). *National strategy for tourism development to 2020* (Vision 2030). Retrieved April 2019, from www.vietnamtourism.gov.vn. (In Vietnamese).

Government of Vietnam (GOV). (2008). *Vietnam biodiversity law.* Retrieved May 2019, from http://vanban.chinhphu.vn/. (In Vietnamese).

Ha, T. B. (2018). *Vietnam administration of seas and Islands, marine protected areas: Orientation for widening and solution for collaborative management.* Retrieved May 2019, from http://vea.gov.vn. (In Vietnamese).

Hobson, P. J., Heung, V., & Chon, K. S. (1994). Vietnam's tourism industry: Can it be kept afloat? *Cornell Hotel and Restaurant Administration Quarterly*, *35*(5), 42–48.

Huxford, K. M. L. (2010). *Tracing tourism translations: Opening the black box of development assistance in community-based tourism in Vietnam* (Master's Thesis, University of Canterbury, Canterbury). Retrieved May 2019, from http://ir.canterbury.ac.nz.

Huynh, B. T. (2011). *The Cai Rang floating market, Vietnam: Towards PPT?* (Master's thesis, Auckland University of Technology, Auckland).

International Centre for Environmental Management (ICEM). (2003). *Vietnam national report on protected areas and development* (Review of Protected Areas and Development in the Lower Mekong River Region). Queensland: Indooroopilly.

International Centre for Environmental Management (ICEM). (2014). *Protected areas and development: lessons from Vietnam.* Retrieved from, http://www.mekong-protected-areas.org/vietnam/docs/vietnam_lessons.pdf

Lanh, V. L., & Truong, X. B. (2018). *Ecotourism in Vietnam's national parks and nature reserves: Status, challenges and solutions.* Retrieved June 2023, from http://vnppa.org/du-lich-sinh-thai-tai-cac-vuon-quoc-gia-va-khu-bao-ton-thien-nhien-viet-nam-tiem-nang-thach-thuc-va-giai-phap.html. (In Vietnamese).

Ly, T. P., & Xiao, H. (2016). The choice of a park management model: A case study of Phong Nha-Ke Bang National Park in Vietnam. *Tourism Management Perspectives*, *17*, 1–15.

Ministry of Agriculture and Rural Development (MARD), Vietnam Administration of Forestry. (2017). *Report of inspection results of business activities of ecotourism services in Vietnamese National Parks and Nature Reserve.* Hanoi: MARD. (In Vietnamese).

Mok, C., & Lam, T. (2000). Vietnam's tourism industry: Its potential and challenges. In K. S. Chon (Ed.), *Tourism in Southeast Asia—A new direction* (pp. 157–164). New York: Haworth Hospitality Press.

Nguyen, T. K. D. (2002). Sustainable tourism development in Vietnam. In T. Hundloe (Ed.), *Linking green productivity to ecotourism: Experiences in the Asia-Pacific region* (pp. 249–263). Tokyo: Asian Productivity Organization.

Nguyen, K. D., Cu, C. L., Vu, H. C., & Tran, M. (2007). Pro-poor tourism in the GMS: Vietnam case study. *Pro-poor tourism in the Greater Mekong sub-region*, 181–217.

Pham, H. L. (2017). *Orientation for sustainable eco-tourism development in National Parks and Nature Reserve*, Vol. 3. Vietnam Environment Administration Magazine. Retrieved May 2019, from http://tapchimoitruong.vn/. (In Vietnamese).

Sandbrook, C. G. (2010). Local economic impact of different forms of nature-based tourism. *Conservation Letters*, *3*(1), 21–28.

Tin, H. C., Uyen, N. T., Hieu, D. V., Ni, T. N., Tu, N. H., & Saizen, I. (2020). Decadal dynamics and challenges for seagrass beds management in Cu Lao Cham marine protected area, central Vietnam. *Environment, Development and Sustainability*, *22*, 7639–7660.

Tran, D. T. (2005). *Introduction to tourism*. Hanoi: VNU Publishers. (In Vietnamese).

Truong, V. D. (2014). *Tourism and poverty alleviation: A case study of Sapa, Vietnam* (Doctoral thesis, University of Canterbury, Canterbury).

United Nations Development Programme (UNDP). (2009). *A mapping exercise – Poverty reduction programme and policies in Vietnam*. Hanoi, Vietnam: UNDP.

United Nations Development Programme (UNDP). (2018). *Human development index*. Retrieved May 2019, from http://hdr.undp.org/en/countries/profiles/VNM.

Vietnam National Administration of Tourism (VNAT). (2005). *Forty-five years of Vietnam's tourism sector*. Retrieved May 2019, from www.vietnamtourism.gov.vn. (In Vietnamese).

Vietnam National Administration of Tourism (VNAT). (2018). *Tourism statistics*. Retrieved May 2019, from vietnamtourism.gov.vn

World Bank (WB). (2018). *Climbing the ladder. Poverty reduction and shared prosperity in Vietnam* (Update Report). Geneva: The World Bank.

Part II

An introduction to case studies

3 Yakushima

3.1 Overview of Yakushima

Yakushima is an island located approximately 60 km off the coast of Kagoshima Prefecture in southern Kyushu, with an area of about 505 km² and a circumference of 132 km. The island is renowned for its diverse biogeographic boundary between the tropical and temperate regions, encompassing all the climatic zones of the Japanese islands from Hokkaido to Kyushu (Okano & Matsuda, 2013). Around 90% of Yakushima's total land area is forest, which was a sacred place for the islanders and a vital resource for the people's livelihoods in the past (Song & Kuwahara, 2016). Historically, the island's cedar, called Yakusugi, was found in the deep mountain areas and was revered by local people. Yakusugi was used as an annual tribute to the Satsuma domain from the mid-18th century, and thus, full-scale logging began on the island (Kanetaka & Funck, 2012).

During the Meiji era (late-19th century) until after the Second World War, Yakushima experienced significant development in the forestry industry during times of high economic growth. As a result, deforestation peaked in the 1960s. However, in the 1970s, there was a shift in governmental forestry policy with a focus toward the conservation of nature, particularly the protection of Yakusugi (Song & Kuwahara, 2016). Thus, this could be considered as a shift in the use of forest resources for tourism purposes (Kanetaka & Funck, 2012). In 1993, 20% of the island was registered as the first Natural World Heritage Site by UNESCO in Japan. Yakushima National Park (NP), which covers the area of the NWHS, was established as part of Kirishima-Yakushima NP in 1964 but became an independent park in 2012. Nowadays, Yakusugi trees are among the main attractions on Yakushima Island.

Presently, the island comprises 24 settlements with two main towns. The largest town, Miyanoura, is situated in the north, followed by Anbo in the east (see Figure 3.1). In 1960, Yakushima's population reached its peak at 24,010 inhabitants, after which it declined to 13,860 in 1995, and has since stabilized at around 13,000 inhabitants (as of 2015). The cessation of large-scale forestry is one of the reasons for the declining population on Yakushima Island. Compared to other remote islands in Japan, the population on Yakushima Island is relatively stable.

DOI: 10.4324/9781003496748-6

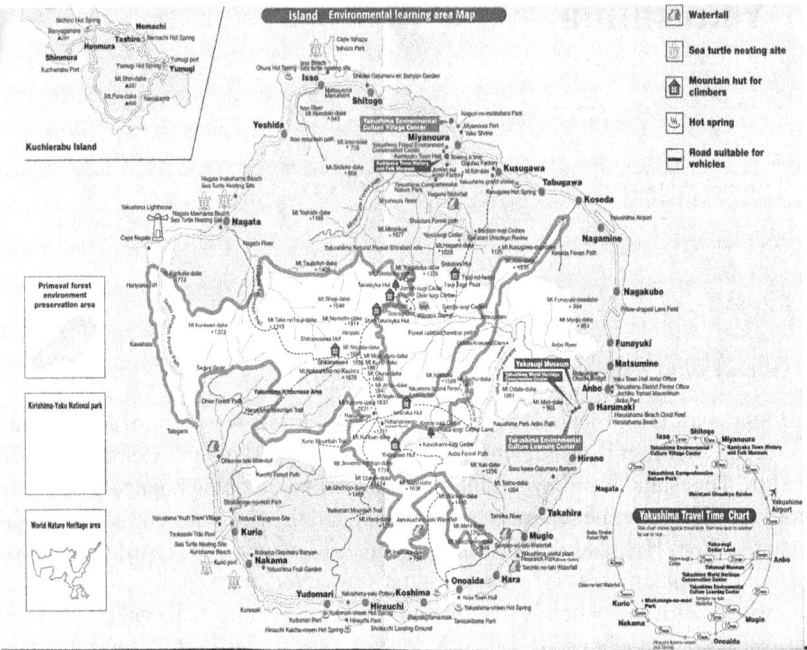

Figure 3.1 Map of Yakushima.

Source: Yakushima Town (2016a)

3.2   Tourism development

Tourism development in Yakushima began in the 1970s, with the preparation of access roads, parking lots, hiking trails and signs to provide information about the island's natural features in the two designated forest areas of the national park, intended for serving tourists (D' Hauteserre & Funck, 2016). The number of visitors to the island increased from 46,000 in 1969 to 122,000 in 1988 (Yakushima Town, 2015). Another significant increase in tourism occurred in 1989 when the first high-speed boat connection between the Kyushu mainland and the island was established. Since then, the island's designation as an NWHS has attracted a large number of tourists, resulting in a rapid increase in visitor arrivals for nearly two decades. From 1995 to 2000, an average of 268,000 people visited the island annually (Hiyoshi, 2002), and from 2006 to 2010, the number of visitors increased to an average of 359,000 per year (Forbes, 2012). The highest annual visitor arrivals were recorded in 2007, with 406,387 visitors, and the numbers have since fallen to 295,972 in 2017 (see Figure 3.2).

Although the number of tourists visiting Yakushima has significantly grown, most of them tend to visit the tourism spots in the mountainous area. Consequently, the number of eco-tour guides has also increased dramatically from 20 guides in 1992 to 164 guides in 2012 (Funck & Cooper, 2013; Okano & Matsuda, 2013).

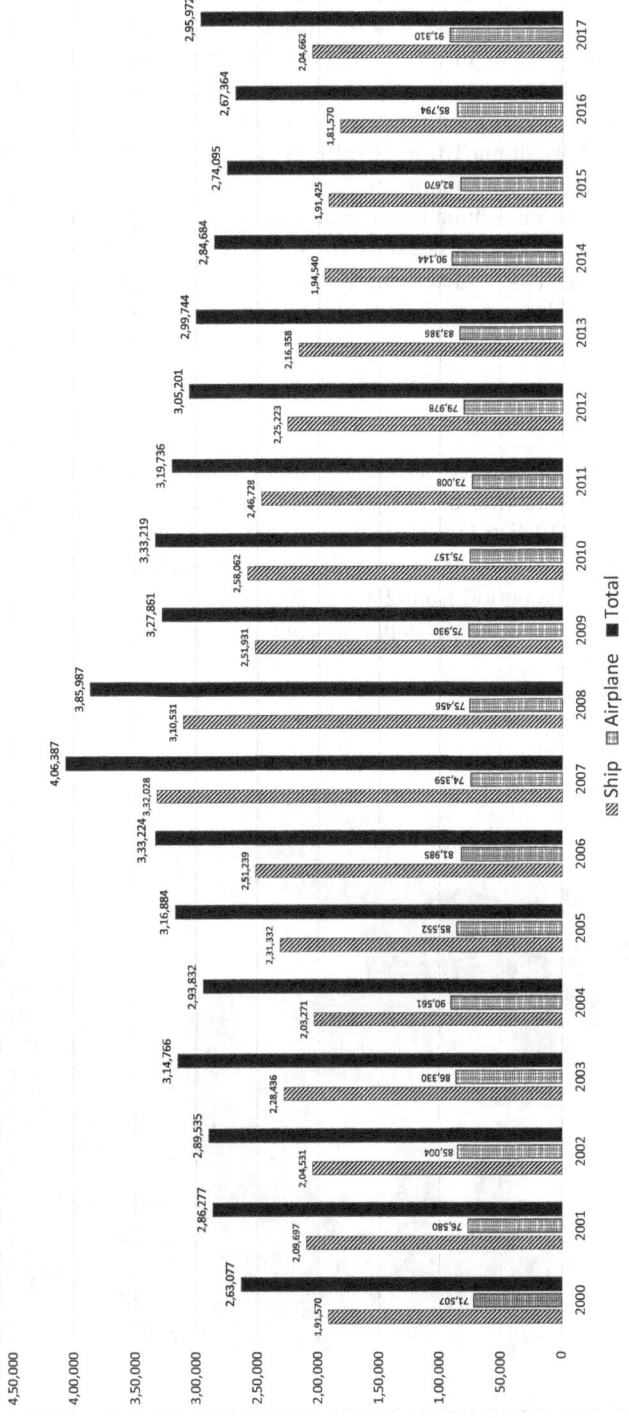

Figure 3.2 Yakushima's tourist arrivals 2000–2017 (unit: person).

Source: Yakushima Town, Statistics on Yakushima, (2018)

According to Kanetaka and Funck (2012), around 80% of guides are non-native migrants from other areas of Japan, and the majority work part-time. As for the number of local enterprises on the island registered as members of the Yakushima Tourism Association, there are currently about 101 accommodation facilities, 31 bento shops and restaurants, 15 car rental companies (with 21 offices) and 95 registered tour guides (Yakushima Town, 2016b). Figures 3.3–3.8 provide some pictures illustrating local tourism enterprises' facilities.

Although tourism in Yakushima is seasonal, running from March until November, the industry brings benefits not only for the local people but also for new inhabitants. This sector has attracted many migrants from the mainland who are involved in providing diverse tourism services, such as tour guides or operating hotels and restaurants.

3.3   The Yakushima economy

Historically, the forestry industry played an important role on Yakushima. However, today the major industries on the island are agriculture, forestry, fishery and tourism. In terms of GDP distribution among economic industries, the tertiary sector, including tourism services, has been a significant contributor to the economy of Yakushima during the period from 2004 to 2014 (as shown in Table 3.1). The data on Table 3.1 also show the per capita income of the population on the island during the same period.

Figure 3.3 An accommodation located in Anbo town.

Figure 3.4 An accommodation located in Hirauchi village.

Figure 3.5 A souvenir shop located in Kusugawa village.

Figure 3.6 A tea shop located in Nagamine village.

Figure 3.7 A souvenir shop located in Miyanoura town.

Figure 3.8 A car rental office located in Koseda village.

In the past, administratively, Yakushima Island had been divided into two areas: Kamiyaku town in the northern half and Yaku town in the southern half until October 2007. Today, the two former towns have consolidated into Yakushima Town. During the period from 1970 to 2015, there were substantial changes in the trends of the population working in different industries, as well as in the total working-age population in Yakushima (as shown in Table 3.2). The tertiary industry has risen to become the most significant contributor to Yakushima's economy.

As discussed earlier, absolute poverty is not a major problem in Japan or in Yakushima. However, income disparity between remote islands and other regions is a pressing issue in Japan. To help financially support these isolated islands and overcome economic challenges, the Japanese government formulated the "Remote Islands Development Act" in 1953 (Kuwahara, 2012). As noted earlier, the Yakushima district belongs to Kagoshima Prefecture, where the relative poverty rate is relatively high compared to the national rate in Japan (as shown in Table 3.3).

Although tourism is an important sector today, its growth has led to a disparity between villages, particularly between tourist and non-tourist villages on the island. For example, according to data provided in Table 3.4, 53.1% of Yakushima's total residents live outside the two main towns of Miyanoura and Anbo. However, only 36.7% of accommodations and 25.9% of bento shops and restaurants are located in those areas. Additionally, as mentioned earlier, the tourism

Table 3.1 Trends in GDP in Yakushima 2004–2014

Year / Sector	2004	2005	2006	2007	2008	2009	2010	2011	2012	2013	2014
Total production	40,835	45,021	45,442	45,010	43,084	41,478	45,482	44,455	41,460	42,721	43,811
Primary	1,724	1,524	1,447	1,263	1,166	904	930	1,061	1,091	1,039	2,054
Secondary	9,546	11,705	11,130	11,209	10,023	9,149	13,127	11,495	8,465	9,802	9,463
Tertiary	30,990	33,676	34,791	32,297	31,595	31,357	31,163	31,588	31,621	31,541	31,853
Income per capita (thousand yen)	1,976	2,063	2,068	2,192	2,098	2,076	2,265	2,240	2,144	2,128	2,106*

Source: Yakushima Town, Statistics on Yakushima (2010, 2011, 2012, 2013, 2014, 2015, 2016b)

Note. Unit: million JP¥

*This is equivalent to about US$20,000

Table 3.2 Trends in population working by industry

Year Sector	1970*	1980*	1990	1995	2000	2005	2010	2015
Primary	1,472	558	1,489	1,283	973	938	882	771
	39%	16%	24%	19%	15%	14%	13%	12%
Secondary	1,027	1,178	1,658	1,683	1,556	1,172	996	995
	27%	34%	26%	25%	23%	18%	15%	15%
Tertiary	1,303	1,696	3,188	3,698	4,150	4,526	4,779	4,712
	34%	50%	50%	56%	62%	68%	72%	73%
Total	3,802	3,432	6,335	6,664	6,679	6,636	6,657	6,478

Source: Yakushima Town (2006, 2010, 2011, 2012, 2013, 2014, 2015, 2016b)

Note: Data of Kamiyaku Town only

Table 3.3 Trends in poverty rate between Kagoshima and nationwide

Year	Nationwide	Kagoshima
1992	9.2%	20.8%
1997	10.1%	18.2%
2002	14.6%	23.0%
2007	14.4%	21.5%
2012	18.3%	24.3%

Source: Tomuro (2016)

Table 3.4 The distribution of some characteristics by village in Yakushima in 2015

Village	% of residents	% of households	% of accommodation*	% of bento shops & restaurants*
Miyanoura	25.2	23.5	33.6	51.6
Anbo	21.7	21.7	29.7	22.5
Other villages	53.1	55.2	36.7	25.9

Source: Created by author based on Statistics Bureau, Ministry of Internal Affairs and Communications, Japan Population Census, 2015 *Source: Yakushima Tourism Association (Yakushima Town, 2016a)

industry in Yakushima has attracted many migrants to the island, making people who work in tourism-related jobs one of the largest workforces on the island. The reasons for this are not only because of the attractiveness of the tourism industry, but also due to the aging and depopulation of the island, which is a common problem in many rural areas in Japan, leading to the demand for human resources from other regions. As a result, economic development on the island has become unequal.

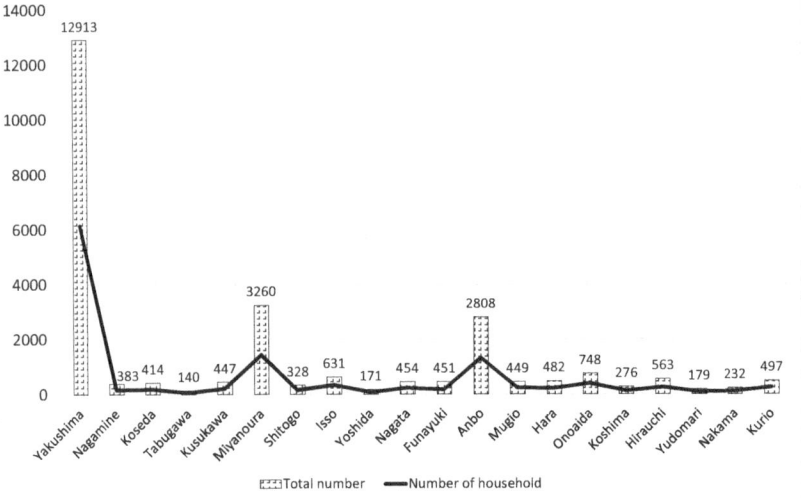

Figure 3.9 Population and household differences by village in Yakushima 2015.

Source: Japan Population Census (2015)

Furthermore, Figure 3.9 provides detailed information on population and household differentiation among villages on Yakushima Island, indicating that there are significant inequalities in the distribution of the population on the island.

3.4   Regulations of NWHS in Yakushima

The nature conservation system in Yakushima is highly complex, as it consists of four different types of nature protection systems that have distinct goals, areas, administration and regulations, as outlined by Tokumaru (2003). These four systems are the "Wilderness Area", "National Park", "Forest Ecosystem Reserve" and "Natural Monuments". Table 3.5 provides an overview of each system.

In summary, the regulations listed in Table 3.5 have had an impact on the lives of the local people on Yakushima island. In the past, the people's traditional livelihoods were strongly dependent on agriculture and forestry industries, which have been affected by the establishment of nature protection systems. Due to the restrictive regulations, such as a ban on forestry operations without government approval, the number of people working in primary industries has decreased, as shown in Tables 3.1 and 3.2. Consequently, the regulations have led to a significant change in the local people's livelihoods on Yakushima island. Since the 1970s, nature-based tourism has emerged as a new industry on the island.

Table 3.5 Outline of the policies surrounding the Yakushima World Heritage Area

Systems name	Purpose and requirements	Zone (Area)	Restrictions	Notes
Wilderness Area (Yakushima Wilderness Area)	Natural areas preserve their primeval state, have not been affected by human activity, and need protection.	1219ha	The following activities require the approval of the Minister of the Environment (Article 17): (1) constructing, reconstructing, or extending buildings, (2) altering the landscape, (3) mining and quarrying, (4) reclamation, (5) altering the water level or water volume in rivers, etc., (6) cutting trees or bamboo (7) collecting fallen leaves or branches, (8) planting trees or bamboo, (9) capturing animals, (10) grazing livestock, (11) lighting fires, (12) accumulating things, (13) using vehicles or horses in designated zones(14) leaving waste, (15) planting plants	Nature Conservation Law
National Park (Kirishima-yaku National Park)	A park contributes to human health, recreation, and enlightenment by protecting scenic beauty and promoting its use	Special Zone (12,005 ha)	The following activities require the approval of the Minister of the Environment (Article 17): (1) constructing, reconstructing, or extending buildings, (2) cutting trees or bamboo, (3) mining and quarrying, (4) altering water level or water volume in rivers, etc., (5) discharging waste water or sewage into designated lakes, marshes, rivers, etc., (6) advertising, (7) reclamation, (8) altering the, (9) collecting designated plants, (10) changing the color of, etc., (11) using vehicles or horses in designated zones	Natural Parks Law

(Continued)

Table 3.5 (Continued)

Systems name	Purpose and requirements	Zone (Area)	Restrictions	Notes
		Special Protection Zone (6,733 ha)	In addition to (1) through (8), (10), and (11) listed above, the following activities require the approval of the Minister of the Environment (Article 18): (12) damaging trees and bamboo, (13) planting trees and bamboo, (14) grazing livestock, (15) dumping, (16) lighting fires, (17) collecting fallen leaves and branches, (18) capturing animals.	"On the rearrangement and expansion of Conservation of Forests" (Order by Director General, of the Forestry Agency, April 1989)
Forest Ecosystem Reserve (Yakushima Forest Ecosystem Reserve)	Through the preservation of primitive forests, Conservation Forests contribute to the maintenance of the natural forest ecosystem, the protection of plants and animals, the preservation of the gene pool, the advancement of forestry management and techniques, and academic research.	Preservation Area (9,600 ha)	Principally, this area will be left to natural succession without human disturbance (emergency measures caused by natural disasters; monitoring, academic research and other activities that contribute to the larger public welfare; and other legal measures are excluded by this restriction).	
		Conservation and Utilization Area: (5,585 ha)	No forestry operations are permitted for the purpose timber production (if there is an artificial forest, multiple forest management may be applicable). Educational and recreational use that does not accompany large-scale development is permissible.	

Natural Monuments	Monuments are animals, plants, and geological minerals of great academic interest in Japan.	Special Natural Monuments: Yakushima Virgin Cedar Forest; Natural Monument: Ryukyu robin (Erithacus komodori), Japanese wood pigeon (Columba janthica), (Turdus celaenops), (Phylloscopus ijimae)	The Commissioner for Cultural Affairs must approve activities that may impact the preservation of the status quo of designated species. (Article 80)	Law on the Protection of Cultural Properties

Source: Ministry of the Environment, Government of Japan, 2019

Reference list

D' Hauteserre, A. M., & Funck, C. (2016). Innovation in island ecotourism in different contexts: Yakushima (Japan) and Tahiti and its Islands. *Island Studies Journal*, *11*(1).

Forbes, G. (2012). Yakushima: Balancing long-term environmental sustainability and economic opportunity. *Kagoshima Immaculate Heart College Research Bulletin*, *42*, 35–49.

Funck, C., & Cooper, M. (2013). *Japanese tourism: Spaces, places and structures*. New York: Berghahn.

Hiyoshi, M. (2002). How the island should consider the increase of tourists to nearly 300,000. *Seimei no Shima*, *17*(2), 97–104. (In Japanese).

Kanetaka, F., & Funck, C. (2012). The development of the tourism industry in Yakushima and its spatial characteristics. *Studies in Environmental Sciences*, *6*, 65–82. (In Japanese).

Kuwahara, S. (2012). The development of small islands in Japan: An historical perspective. *Journal of Marine and Island Cultures, 1*, 38–45.

Okano, T., & Matsuda, H. (2013). Bio-cultural diversity of Yakushima Island: Mountains, beaches, and sea. *Journal of Marine and Island Cultures*, *2*(2), 69–77.

Song, D., & Kuwahara, S. (2016). Ecotourism and world natural heritage: Its influence on islands in Japan. *Journal of Marine and Island Cultures*, *5*, 36–46.

Tokumaru, H. (2003). *Nature conservation on Yakushima island: Kagoshima prefecture's efforts*. Kagoshima: Nature Conservation Division, Kagoshima Prefectural Government.

Tomuro, K. (2016). *Trends observed in poverty rates, working poor rates, child poverty rates and take-up rates of public assistance across 47 prefectures in Japan* (Yamagata University Faculty of Humanities Research Annual Report No. 13), 33–53 (In Japanese).

Yakushima Town. (2006). *Statistics on Kamiyaku Town*. Retrieved from www.town.yakushima.kagoshima.jp. (In Japanese).

Yakushima Town. (2010). *Statistics on Yakushima*. Retrieved from www.town.yakushima.kagoshima.jp. (In Japanese).

Yakushima Town. (2011). *Statistics on Yakushima*. Retrieved from www.town.yakushima.kagoshima.jp. (In Japanese).

Yakushima Town. (2012). *Statistics on Yakushima*. Retrieved from www.town.yakushima.kagoshima.jp. (In Japanese).

Yakushima Town. (2013). *Statistics on Yakushima*. Retrieved from www.town.yakushima.kagoshima.jp. (In Japanese).

Yakushima Town. (2014). *Statistics on Yakushima*. Retrieved from www.town.yakushima.kagoshima.jp. (In Japanese).

Yakushima Town. (2015). *Statistics on Yakushima*. Retrieved from www.town.yakushima.kagoshima.jp. (In Japanese).

Yakushima Town. (2016a). *Yakushima tourism association*. Retrieved from http://yakukan.jp/ (In Japanese).

Yakushima Town. (2016b). *Statistics on Yakushima*. Retrieved from www.town.yakushima.kagoshima.jp. (In Japanese).

Yakushima Town. (2017). *Statistics on Yakushima*. Retrieved from www.town.yakushima.kagoshima.jp. (In Japanese).

Yakushima Town. (2018). *Official certified guide of Yakushima*. Retrieved from www.yakushima-eco.com/ (In Japanese).

4 Cu Lao Cham

4.1 Overview of Cu Lao Cham

Cu Lao Cham Island, also known as Tan Hiep Commune, consists of eight islets with a total area of approximately 15 km². It is located in the central province of Quang Nam, approximately 18 km offshore from Hoi An City (see Figure 4.1). There are four villages on the island, consisting of about 2,500 people and 600 households. In late 2005, the Cu Lao Cham marine protected area (MPA) was established under the authority of the Quang Nam Provincial People's Committee (QNPPC) and was the outcome of the Cu Lao Cham MPA project, a livelihood support program funded by the Danish International Development Agency (DANIDA) (Brown, 2011). The primary objectives for establishing the MPA were to preserve the biodiversity value by protecting the fish and their ecosystem, as well as to improve the local livelihoods of Cu Lao Cham Island. The Cu Lao Cham MPA covers an area of 6,710 hectares and includes both an island nature reserve and protected marine waters (Nguyen, 2010). Due to its rich biodiversity value, Cu Lao Cham was designated as a UNESCO World Biosphere Reserve (WBR) in 2009 (Nguyen, 2010).

4.2 Tourism development

Residents of Cu Lao Cham have witnessed significant changes, as their home island has transitioned from a small fishing community to a popular tourist destination (Brown, 2011). Before 2005, fishing was the main economic activity on the island, with 90% of Cu Lao Cham's household heads employed in the fishing industry (Tri, 2007). In 2005, tourism began to develop, with approximately 5,000 tourist arrivals, although activities were restricted to day trips organized by tour groups (Brown, 2011). The number of visitors increased rapidly, reaching an annual total of 407,315 in 2017, but decreased slightly to 399,682 in 2018 (see Figure 4.2). Tourists are attracted to the island not only for the beauty of the beaches, coral reefs and scenery but also for the rich historical and cultural heritage of the local communities, reflected in their pagodas, old temples and traditional folklore. Common tourist activities on Cu Lao Cham include sightseeing tours by motorbike taxi, swimming, snorkelling and scuba diving. Tourism on Cu Lao Cham is seasonal,

DOI: 10.4324/9781003496748-7

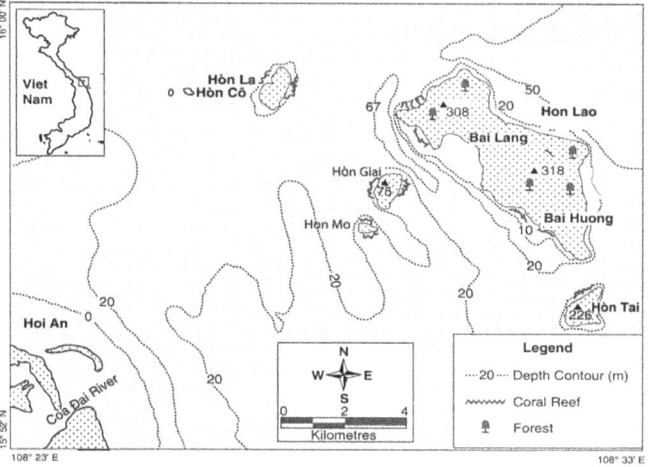

Figure 4.1 Location of Cu Lao Cham Island.

Source: Ashton (2004)

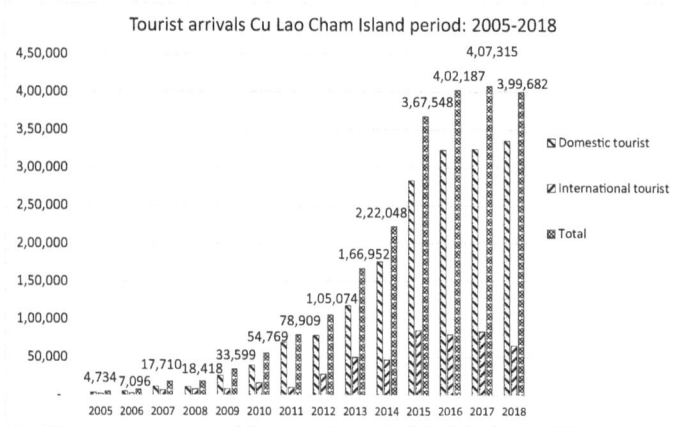

Figure 4.2 Tourist arrivals Cu Lao Cham 2005–2018 (unit: person).

Source: THPC (2018)

typically running from March until August; the rest of the year is considered off-season due to the frequent occurrence of typhoons.

4.3   The Cu Lao Cham economy

Cu Lao Cham witnessed a significant change in its economic sector from 2011 to 2015. During this period, the proportion of agroforestry and fishing decreased from 69.89% in 2012 to 20.27% in 2015, while the proportion of tourism and services increased from 30.11% in 2012 to 70.83% in 2015, making it a spearhead economic sector on the island. In 2011, Cu Lao Cham's production value was recorded at VND 33 billion (US$1.6 million). However, in 2015, the total production value increased by 3.3 times, with a recorded value of VND 120 billion (US$6.0 million) (Hoi An People's Committee, 2016).

The fishery industry has traditionally been a significant source of livelihood for many local people in Cu Lao Cham. However, due to the conservation regulations imposed when establishing the Cu Lao Cham MPA, tourism has emerged as an important alternative livelihood for the locals (Chu, 2014; QNPPC, 2017). According to Tan Hiep People's Committee (THPC), the number of households and total labor involved in the tertiary industry, primarily tourism, has increased rapidly from 2008 to 2016 (see Figure 4.3).

Currently, approximately 70% of people on the island have shifted from fishing to tourism services (Vietnam News (VNS), 2018). Over the past decade, there has been a significant increase in the number of households involved in tourism services (see Figure 4.4). Tourism has not only substantially contributed to the islanders' income per capita but also helped alleviate poverty among the poor

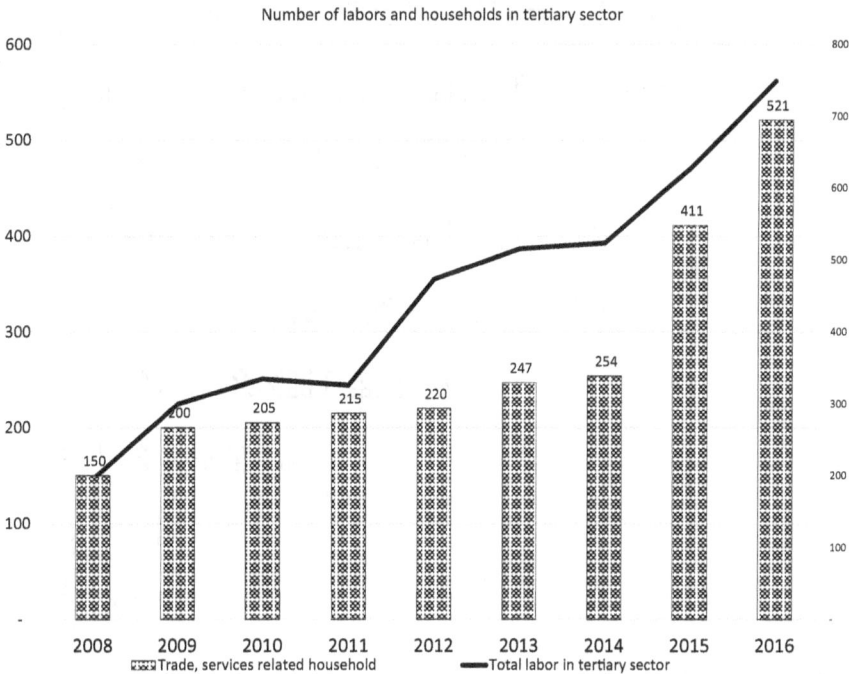

Figure 4.3 The changes in labor and households in the tertiary sector in Cu Lao Cham 2008–2016.

Source: THPC (2018)

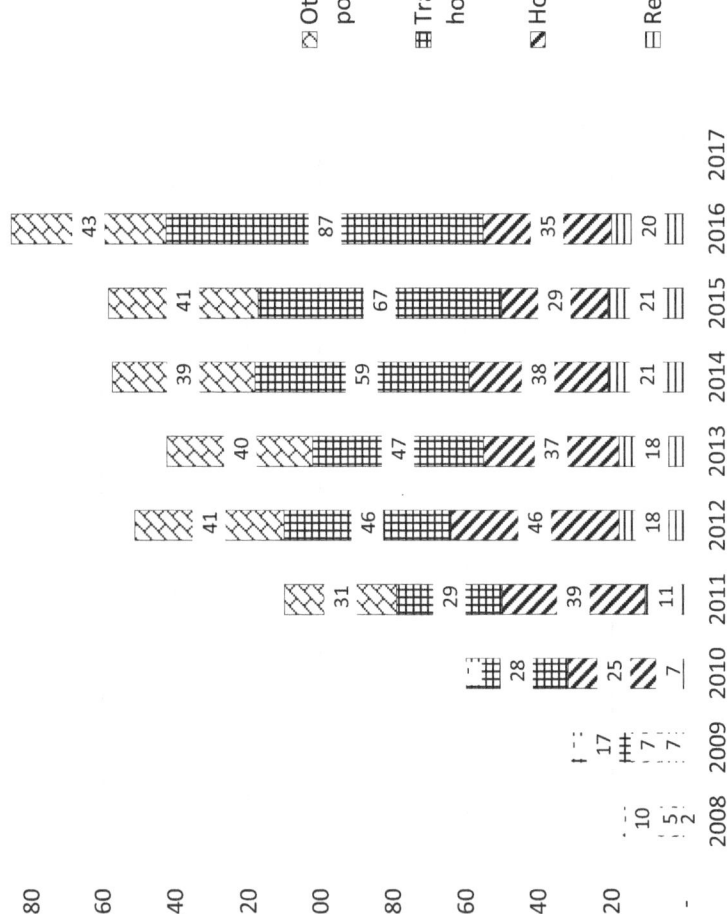

Figure 4.4 Local tourism businesses in Cu Lao Cham from 2008 to 2016.

Source: THPC (2018)

and near-poor households. According to the Hoi An People's Committee (2016), the income per capita of the people of Cu Lao Cham was around VND12 million (US$600) in 2010, and this increased to VND28.6 million (US$1,400) in 2015. By the end of 2015, the proportion of poor and near-poor households had decreased significantly on Cu Lao Cham.

VNS (2018) reports that while tourism has brought economic benefits to Cu Lao Cham, local people only receive one-third of the tourism revenues, with most profits going to businesses from the mainland. The impact of tourism on benefits distribution on the island will be discussed further in Chapter 6. Figures 4.5–4.12

Figure 4.5 A signboard of homestay at Bai Ong village.

Figure 4.6 A signboard of homestay at Bai Lang village.

Figure 4.7 Souvenir shops at the tourist market.

Figure 4.8 Dried seafood sellers along the commune road.

Figure 4.9 Tourist pier in Cu Lao Cham.

Figure 4.10 Tour operator coming to Cu Lao Cham.

Figure 4.11 Motorbike taxi service.

Figure 4.12 A signboard of coral watching service.

provide some pictures that illustrate the local tourism enterprises on Cu Lao Cham Island.

4.4 Regulations of MPA in Cu Lao Cham

In Cu Lao Cham, a zoning system was implemented when establishing the MPA. According to QNPPC (2005), the zoning management of the Cu Lao Cham MPA was divided into three categories: core zone, ecological rehabilitation zone and controlled development zone (see Figure 4.13). The controlled development zone comprises the tourism development zone, community development zone and reasonable fishing zone (see Table 4.1). According to the Management Board of Cu Lao Cham MPA (2006), 270 of the island's 500 households were involved in some way in discussions about the draft zoning plan and regulations. Those who were fishing in the core zone and near-shore coral reefs were most adversely affected by the zoning plan and regulations. Regulations of each zone are formed according to a range of permitted activities, some of which are prohibited within the entire MPA area, with the strictest controls evident in the core zone. The prohibited activities are provided in Table 4.2.

Despite the zoning system and regulations, local people have a more flexible view of what constitutes acceptable use of natural resources than what is specified by the Management Board of Cu Lao Cham MPA (Brown, 2011). However,

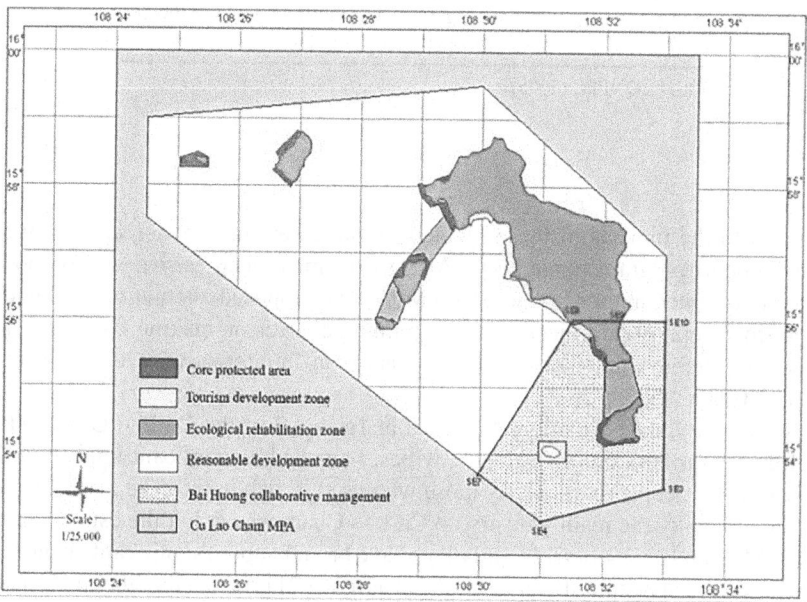

Figure 4.13 Regulated zones of protection of Cu Lao Cham MPA.

Source: Chu (2009)

Table 4.1 Zones in the Cu Lao Cham MPA management structure

MPA zone	Zone description
Core Zone	totally conserved, managed and carefully protected, maximum restriction of use to prevent negative impacts on habitats, may be used for scientific research, training and education
Ecological Rehabilitation Zone	managed, protected and well organized activities for recovering ecological habitats, biodiversity and natural marine resources in order to economically benefit communities
Controlled Development Zone	
Tourism Development Zone	tourism activities which are able to generate income for the local people, and controlled by the Management Board of MPA and include: scuba diving, coral reefs viewing by glass bottom boat, surfing, sailing, swimming, research, education, training, community entertainment
Community Development Zone	located on lands where people are living in Bai Lang, Thon Cam, Bai Ong and Bai Huong villages of Cu Lao Cham island
Reasonable Fishing Zone	reserved for organizing reasonable marine resources exploitation, developing relevant activities (fishing, aquaculture and other suitable gears) in order to increase income, improve living standards and alternative income generation for MPA communities. The reasonable fishing zone surrounds the extremely protected zone, the ecological rehabilitation zone and the controlled development zone

Source: Brown (2011); QNPPC (2005)

to adapt to the impacts of these regulations on fishers in the core zone, a livelihood support program funded by DANIDA, as mentioned earlier, was proposed and implemented in the Cu Lao Cham MPA. The proposed alternative livelihoods included several pilot projects, such as training courses on making fish sauce and dried fish, before implementing tourism as the main alternative livelihood in the Cu Lao Cham MPA.

In summary, the regulations specified in Table 4.2 of the Cu Lao Cham MPA, such as the prohibition of fishing activities, have had a significant impact on the lives of local people, particularly those who used to fish in the core zone. Fishing was traditionally the main industry on Cu Lao Cham island, but the establishment of the MPA and its restrictive regulations have brought about significant changes in the lives of local people. Tourism emerged as a new alternative livelihood for many locals, as noted earlier, and since 2005, this sector has become a vital contributor to the local economy.

Table 4.2 Prohibited activities within the Cu Lao Cham MPA and associated zones

Zone	Prohibited activities
Entire Cu Lao Cham MPA	a) Activities disturbing the environment, landscapes; destroying substratum rock, coral reefs, flora and other ecosystems; negatively impact on marine species' communities, habitats, breeding and growth areas. b) Fishing activities by dynamite, chemicals, electricity, poison and other destructive methods. c) Hunting of fauna and flora species which are named on the protected list. d) Exploiting activities which are named on the banned list including seasonal ban, except for in cases of research purposes permitted by the Government; Exploitation of marine animals which body size smaller than specified sizes, except for allowed catches for aquaculture purposes; e) Industrial scale aquaculture; f) Any kind of mining; g) Activities that cause beach erosion around islands; h) Activities that illegally occupy, convert land, or water use; i) Activities that introduce non-endemic flora or fauna species that might cause damage to the environment, natural ecosystems or biodiversity of the MPA; k) Activities that pollute the environment including noise, vibration where the intensity is greater than permitted limit.
Ecological Rehabilitation Zone	In addition to all restrictions listed above: a) Construction, housing, anchoring in coral reef areas; b) Any kinds of exploitation of forestry or aquatic products.
Core Zone	In addition to all restrictions listed above, the following activities are also prohibited: • Collecting mineral specimen, coral, wild animals, aquatic fauna and flora, microbiological samples; • Any kinds of visiting or excursion, touring, swimming, snorkeling, diving.

Sources: Brown (2011); QNPPC (2005)

Reference list

Ashton, E. C. (2004). *Utilization of aquatic biodiversity by the local communities living in and around the Cham islands*. Hoi An: CLC MPA Project.

Brown, P. (2011). Livelihood change around marine protected areas in Vietnam: A case study of Cu Lao Cham. In Canada Research Chair in Asian Studies (Ed.), *ChATSEA working paper no. 16*. Montreal: Universite de Montreal.

Chu, M. T. (2009). *Cooperation between marine protected areas and integrated coastal zone management in Quang Nam Province, Vietnam*. Paper presented to Workshop on Networking of Protected Areas: Benefits, Good Practices, Standards and Next Steps, Manila.

Chu, M. T. (2014). *Opportunities and challenges in the management and preservation of biodiversity value of Cu Lao Cham Biosphere Reserve*. Hoi An. Retrieved October 10, 2018, from www. khusinhquyenculaocham.com.vn. (In Vietnamese).

Hoi An People's Committee. (2016). *Project of sustainable development for Tan Hiep Commune towards preservation of Cu Lao Cham World Biosphere Reserve.* Retrieved April 2019, from http://hoian.gov.vn. (In Vietnamese).

Management Board of Cu Lao Cham MPA. (2006). *Completion report on the Cu Lao Cham MPA project's activities (10/2003-09/2006),* People's Committee of Quang Nam Province, Hoi An.

Nguyen, T. T. N. (2010). *Effectiveness evaluation of a marine protected area in Vietnam— The Cu Lao Cham MPA case study* (Master's Thesis, University of Tromsø, Tromsø).

Quang Nam Provincial People's Committee (QNPPC). (2005). *Decision No. 88/2005/ QĐ-UBND of promulgating the regulation on management of Cu Lao Cham MPA, Quang Nam province.* Retrieved June 2019, from culaochammpa.com.vn

Quang Nam Provincial People's Committee (QNPPC). (2017). *Decision No. 2494/ QĐ-UBND of promulgating the regulation on management and organization of tourism and sports activities in Cu Lao Cham MPA, Hoi An city, Quang Nam province.* Retrieved October 2018, from www.thuvienphapluat.vn

Tan Hiep People's Committee (THPC). (2018). *Personal correspondence with Cu Lao Cham tourism management board.* Hoi An: Tan Hiep People's Committee

Tri, H. M. (2007). *Survey report on income, environmental awareness and livelihoods consultation to affected households.* Hoi An/Hanoi: CLC MPA Management Board/LMPA Program.

Vietnam News (VNS). (2018). *Tourism boom threatens Chàm Island ecosystems.* Retrieved April 2019, from https://vietnamnews.vn/

Part III
Perspectives on tourism and poverty alleviation

5 Perspectives of local businesses on tourism development

The case of Yakushima

5.1 Introduction

The background of tourism development in Yakushima reveals the current reality of socioeconomic aspects on the island, which is provided in Chapter 3. To address the research question of whether or not tourism development creates inequalities on Yakushima, interviews were conducted. This chapter reports and discusses findings from semistructured interviews conducted with local tourism enterprises in Yakushima. The results of the interviews are complemented by field notes taken during the fieldwork. The chapter considers the important issue of local entrepreneurs' perceptions of tourism's contribution to the local economy, its effects on income distribution and its impacts on spatial differences among villages. As mentioned earlier, it is necessary to explore locals' perspectives on tourism's impacts on the local economy, income distribution in the context of nature conservation areas such as Yakushima, an NWHS, to find appropriate strategies not only for tourism development but also for conservation purposes that take into account the voice of local business owners. This chapter starts with a description of the interviewee selection process and details about the interviewees. It then reports on the key themes based on the questions asked. Three key themes were identified that comprise local entrepreneurs' perspectives: (1) perceptions of tourism's impact; (2) perceptions of income distribution; and (3) perceptions of spatial differences in tourism impacts.

5.2 Interview respondents' profiles

As stated, the main collection of primary data for this case study was based on semi-structured interviews conducted with 32 local tourism entrepreneurs in Yakushima. Prior to conducting the interviews, appointments were arranged with tour guides due to their busy schedules. They were chosen based on the list of Official Certified Guides of Yakushima (Yakushima Town, 2018a), according to their registered field of activities. The remaining interviews were conducted randomly throughout the study site, across nine of the 24 villages on Yakushima Island. The majority of interviews took place in Anbo and Miyanoura, which are the primary towns in Yakushima known for having better-developed tourism facilities compared to other

DOI: 10.4324/9781003496748-9

villages on the island. The interviews were predominantly conducted in Japanese, although a few were conducted in English with the non-Japanese participants.

The participants in this study comprised nine tour guides and/or tour operators, six owners and/or staff of accommodation facilities, four owners and/or managers of rental car companies, and 13 proprietors and/or staff of restaurants, souvenir shops and coffee shops. As shown in Table 5.1, of the 32 local entrepreneurs interviewed, 20 were male (62.5%) ranging in age from 27 to 80, and 12 were female (37.5%) between the ages of 35 and 70. The interviewees had an average age of 49. Of the 32 respondents, 31 were Japanese, and one was American, and 20 of them were non-natives, meaning they came from outside Yakushima. All quotes are anonymized. The chapter presents perceptions of tourism's impact first.

Table 5.1 Interview respondents' profiles

No.	Local enterprisers (Name pseudonym)	Gender	Age	Starting year	Location
1	Tour guide A	Male	36	2012	Anbo
2	Tour guide B	Male	41	2012	Miyanoura
3	Tour guide C	Male	55	1993	Anbo
4	Tour guide D	Male	42	2001	Miyanoura
5	Tour guide E	Male	43	1998	Anbo
6	Tour guide F	Male	62	1989	Miyanoura
7	Tour guide G	Female	36	2014	Nagata
8	Tour guide H	Male	37	2003	Isso
9	Tour guide I	Male	37	2013	Anbo
10	Accommodation A	Male	27	2004	Miyanoura
11	Accommodation B	Male	80	1947	Isso
12	Accommodation C	Female	45	2003	Hirauchi
13	Accommodation D	Female	55	1972	Anbo
14	Accommodation E	Male	45	2016	Miyanoura
15	Accommodation F	Female	55	1970	Miyanoura
16	Souvenir shop A	Female	35	2012	Kusugawa
17	Restaurant B	Male	39	1982	Anbo
18	Café shop C	Male	55	2013	Anbo
19	Souvenir shop D	Female	35	2013	Anbo
20	Café shop E	Female	48	2011	Nagata
21	Café shop F	Male	40	2004	Isso
22	Café shop G	Female	70	–	Hirauchi
23	Souvenir shop H	Male	69	1998	Nagamine
24	Restaurant I	Male	50	2006	Kurio
25	Café shop J	Female	70	2012	Kusugawa
26	Souvenir shop K	Female	63	–	Anbo
27	Fruit garden L	Male	70	1983	Kurio
28	Restaurant M	Female	40	–	Anbo
29	Rental car A	Female	50	1995	Koseda
30	Rental car B	Male	40	–	Miyanoura
31	Rental car C	Male	45	2010	Koseda
32	Rental car D	Male	52	2005	Koseda

5.3 Perceptions of tourism's impact

For most interviewees, tourism plays a crucial role in Yakushima Island as it is the primary source of income (29 respondents). Their perceptions of tourism's impact reflect two main points: the general economic impacts and the issues of instability and inequality.

5.3.1 General economic impacts

Of the 31 interviewees asked about the impact of tourism on Yakushima, 30 stated that tourism has a positive impact on the economy in Yakushima Island. One was not able to respond to the question because he had just come to work on the island.

> Obviously, tourism is very important for some of the reasons: First, it is income, lodging. Second, it is the huge contribution on conservation. If you look at the picture 30–40 years ago and compare it with now, Yakushima is one of the very few WNH site that the forest is getting better. (Tour guide G, Anbo)
> Tourism is the main industry on this island. In that meaning, it must be a lot. If there is no tourism on this island, it is really hard to maintain local economy in the future.
> (Tour guide E, Hirauchi)

As previously mentioned, there has been a noticeable shift in the occupational distribution of Yakushima's population from the primary sector to the tertiary sector over the past few decades. This observation was echoed by a tour guide who expressed a similar perception:

> I guess [it] is the most important thing. Since I came, the industrial structure of Yakushima has changed a lot. When I first came, there was still a large proportion of people working in the agriculture, forestry, and fishing industries, but I think the proportion working in the tourism industry, I don't know exactly, has increased now. I can say the number of hotels and guesthouses have already increased tremendously.
> (Tour guide F, 62 years old, Hara)

With regard to immigration on the island, tourism development in Yakushima also attracts new people from outside the island who come to work here. However, due to the decrease of tourists recently, some respondents believed that migrants may also return to their hometowns:

> First of all, on top of the number of local people, the number of tourists increase the population on the island, so more money comes in. For example, tourists spend money on souvenirs, eating at restaurants, staying at guesthouses, and also sightseeing. That means more money flows in through the tourism sector. (Souvenir shop A, Kusugawa)

Tourism in Yakushima has a short history although many people think that Yakushima is only for sightseeing. But, it has only become an important industry since many tourists have come to the island, so towns now do a lot of things related to tourism too. Also, there are a lot of people depending on the tourism industry. However, as the number of tourists decrease, people leave the island. Some of the migrants have also returned to their hometowns.

(Accommodation A, Miyanoura)

One of the positive impacts on Yakushima's economy is that the employment generation for the residents. Of the 31 interviewees asked about the employment generated from tourism, 26 indicated that tourism plays an important role in creating jobs on the island.

Absolutely, if you look at the number of people who work in the hostels, the taxis, the buses, etc.—without tourists none of them could exist. Even people who drive the bus, without tourists it would be nothing. (Tour guide G, Anbo)

For example, there is a high school in Yakushima where every year about 100 people graduate. 95 of them leave the island, but of the remaining five, four are employed in the tourism sector. So, if you think about it in that way, tourism plays a very important role.

(Tour guide C, Anbo)

However, tourism development in Yakushima also has a "life cycle"; like other tourist destinations, the number of tourists was increasing steadily, but recently it has been decreasing. This shift in tourism trends has influenced employment trends, as one tour guide mentioned:

A while ago, an interesting thing happened. A road construction worker in this area became a tour guide. Before I knew it I thought "oh that guy has also become a tour guide", but sometimes, it is the opposite, the tour guide becomes a road construction worker.

(Tour guide C, Anbo)

Three respondents stated that tourism does not create much employment on the island because it is not a big market. Likewise, there is a lack of human resources on the island because of its aging population. The remaining two respondents were not able to answer this question.

Tourism does not create a lot of employments on the island. That is one weak point. Again, it is not a big market, we cannot hire many people, for example in the past I hired two persons. It wasn't easy to pay good salary for them because we have the low season. It is quite difficult to maintain employment throughout the year. (Tour guide E, Hirauchi)

As is often the case with all industries [including the tourism industry], there are only a few people available to employ, so the industries are struggling to find employees. So young people, and women, are coming from outside.

(Tour guide F, Koseda)

The impacts of tourism on Yakushima's economy including employment generation can be seen as the positive effects. But at the same time respondents raised some issues that should be addressed such as the instability of tourism-related jobs and inequality of employment generation within tourism sector and in-outsiders.

5.3.2 Issues of instability and inequality

As previously stated, tourism in Yakushima follows a seasonal pattern, operating from March to November, with no tourism during the remaining months of the year. Hence, the number of potential customers on the island during peak versus low season is very different. Of the 32 respondents, three stated that they do not work during the off-season. Some accommodations and restaurants remain open because they still have working and local customers. Some enterprisers do farming, travelling, marketing or training; some visit their home country, etc.

In the off-season, I do not do anything else, but I reduce my personnel expenses. There are still local people and a few tourists coming, so I keep the store open. (Restaurant B, Anbo)

We give up working in the off-season. Since the rental car work is the only job that we do, I let everyone take a vacation or take a rest, including me. I thought that there were no other choices. So I do not do much.

(Rental car A, Koseda)

The seasonal effect reflects the instability in tourism development on Yakushima, especially in maintaining the stable of enterpriser's income. Some of them have to find the part-time job or stop working during the off-season. Although tourism has created many jobs not only for local people but also for non-local people in Yakushima, it was noted that tourism generates the most jobs for tour guides, especially mountain guides, and those jobs were mainly filled by people from outside of Yakushima:

I guess [tourism creates jobs] only for tour guides. The mountain guides, but the tour guides, many of them are not local people. There are many people who come from outside, not only Japanese but also many foreign tour guides. (Café owner C, Anbo)

About 80 percent of mountain guides come from elsewhere, but when the peak sightseeing period subsides, many people move out. Tourism in Yakushima cannot continue to operate during the off-season, and it is also hard to have regular employment, so people who want to make a contract, with some insurance, for the sake of their children, leave the island. So, many people come to Yakushima, then many people leave, just a few people remain. It has been like that for a long time on this island.

(Café owner F, Isso)

As some respondents stated earlier, it suggests that inequality among tourism enterprisers could exist because most of the tourists visiting the island are interested in hiking and exploring the beauty of nature in the mountainous areas. Therefore, the tour guide is considered one of the profitable jobs in Yakushima's context. This job is mainly occupied by people who moved to Yakushima from outside, thus this tendency may worsen the unevenness in terms of job creation in this site.

The local enterprisers perceived tourism's impact in both positive and negative ways. Although tourism in Yakushima indeed has significantly contributed to the local economy, it creates inequalities due to its unique characteristics like the seasonal effect, and the share of employment by people from the mainland because of human resources in the island are lacking. To understand better how tourism's development influences local businesses' income, the next section explores the perceptions of income distribution.

5.4   Perceptions of income distribution

When asked about income distribution and how that may relate to local tourism enterprises, the 30 interviewees perceived very different points of view. In order to have a better picture of this issue, the perceptions of income distribution is divided into four main points, which include: differences between sectors, differences within sectors, money flows and self-evaluation of respondents' income.

5.4.1   Differences between sectors

Out of the total number of interviewees (30), a majority of them (18) stated that accommodations owners are the ones who earn higher profits compared to other enterprises. Similarly, 15 interviewees (including repeated respondents) regarded tour guides as being among the most benefited individuals. Following these sectors, car rental companies and souvenir shops were identified by four respondents. Among the 30 interviewees, five were unsure about which sectors generated greater profits, while one interviewee mentioned that all sectors experienced equal benefits:

I am certain that the guesthouse businesses can make the most profit. Because you cannot come to this island without staying overnight. (Tour guide A, Anbo)

I think being a tour guide probably earns the most. First of all, you don't need to make any investment, I mean, you don't have to have anything to be a guide. (Accommodation A, Miyanoura)

People definitely need a way to move around by using rental cars or buses. I think you can surely earn some money if people come. Also, as for souvenir shops, Japanese people still have a habit of buying souvenirs. (Rental Car A, Koseda)

Tourism is not a very profitable industry. Customers still come here so that they can manage to run their business, but I think no one makes a lot of money. Their profits are similar although some places have good reputations, and the others do not.

(Tour guide F, Koseda)

Along with sectors with more profits as perceived, restaurants, souvenir shops, and rental cars were considered the least profitable sectors (ten respondents). However, one-third of interviewees (ten) did not know which sectors earn the lowest profits. Interestingly, four of six accommodation owners considered themselves as belonging to the least profitable sector:

Well, small profits? I guess restaurants, other than that, maybe souvenir shops. I mean, those smaller souvenir shops would make a small profit. For those bus companies that don't get involved in tours, they are basically a retail business, so I think they don't make a lot [of profits]. (Restaurant owner M, Anbo)

I think the less profitable ones are probably guesthouses, and maybe also restaurants. At our guesthouse, even when it is full, the money we make per day is 3,000 yen (about US$30) per person, multiplied by thirty per month, which is 90,000 yen (about US$900). But in the case of a tour guide, if there are five customers who take a Jomon Sugi tour, that's already 10,000 yen (about US$100) [per tourist].

(Accommodation A, Miyanoura)

Many tourists who visit Yakushima tend to stay for several nights due to the island's remote location and the abundance of tourist attractions it offers. This extended duration allows them ample opportunities to discover and appreciate the beauty and distinctiveness of Yakushima Island. For instance, a popular shared interest among visitors is embarking on a hiking tour to *Jomon Sugi*. To engage in such activities, tourists require accommodations and often opt to hire tour guides or book package tours. Therefore, it is reasonable that some respondents perceive accommodations and tour guides as more lucrative compared to other entrepreneurs on the island.

5.4.2   *Differences within sectors*

Of the 26 interviewees asked about whether there were existing differences with regard to income or customers within the tourism sector, 24 indicated that there was inequality among local enterprises. The other two respondents did not know the answer:

Perhaps in Yakushima, the mountain guide is the best job, because the income of mountain guides is the largest, the proportion of their customers is also the highest. I think the second highest income is a sea [guide]. The third highest, I think, is a river [guide]. Because the river is so cold, you can only swim there in summer.

(Tour guide A, Anbo)

Within the same tourism businesses, there was the disparity in income distribution. Factors that influence their income, as the respondents mentioned, are the quality of services, the prices, etc. with regard to tour guides, rental cars or restaurants. However, in the case of lodging facilities, accessibility is very important. The accommodations that are located in the two main towns are often chosen by more tourists than other villages.

5.4.3   Where the money flows?

Some respondents were asked about whether tourism-related profits stay on the island or travel off the island. Two respondents stated that money might flow to the big companies outside the island. Some others mentioned that the money stays in Yakushima:

I think the profit stays in the island. But these days, there are many big companies and customers coming with the tour guides from the outside. That means some portion of the profit taken out of the island because they bring their own tour guides, especially with big groups. We cannot say don't come! (Tour guide E, Hirauchi)
    I don't think the money is flowing out from the island. For example, booking.com and Rakuten Travel, etc. earn a lot of money. But basically, money from tourism stays on the island, there is never foreign currency involved . . .

(Accommodation E, Miyanoura)

The perceptions of money flows reveal the problem that tourism benefit not always contributes to local destinations but rather flows somewhere else. This concern is important for considering the income distribution in Yakushima.

5.4.4   Evaluation of respondents' income

Of the 28 interviewees asked about self-evaluation of their income, almost half (11) felt they earned a low income, even compared with the average resident of Yakushima (Tour guide C). Other respondents stated their income places them on an average (seven respondents), relatively good (one respondent), or good (nine respondents) level:

The income is low, even compared with the Yakushima average. In our case, we almost never do anything that we are not interested in, so our earnings suffer from that. For example, the most profitable tour job in Yakushima is the Jomon Sugi guide. Since we don't do it, we earn less. So we suffer from it. (Tour guide C, Anbo)

The income is low if compared with average of Japan. But, again I accepted and I enjoy the life here. That is ok. Many people come to this island to spend their lives here in their ways. If they want to get a lot of money, it is better to work in the big cities.

(Tour guide E, Anbo)

As mentioned earlier, the poverty rate in Kagoshima Prefecture is higher than Japan's national poverty rate. Some interviewees perceived their income as similar to the statistics shown. Although the income was recognized as low or average, many immigrants, including guides in Yakushima, value quality of life higher than earning money. Some of them come to the island to enjoy their lives after retiring.

5.5   Perceptions of spatial differences of tourism's impact

Twenty-nine interviewees were asked whether tourism creates any differences among villages in Yakushima concerning economic effects, employment generation or especially between tourist-villages and non-tourist villages. Twenty-seven stated that there was a gap between Miyanoura, Anbo and other villages. These two main towns were considered most convenient since they are the main gateways and are therefore easily accessible by boats and ferries from the mainland.

The main areas of Anbo and Miyanoura have a lot of lodging facilities. So, after all there are more tourists coming around there, and there is also more money coming in, I guess. Other areas are far away from seaports and airport, the transportation is not good and difficult to access, so money does not come in as much. The difference in transportation makes a big influence. (Tour guide A, Anbo)

Well, I think it depends on the areas. The areas like Miyanoura, Anbo, and Koseda are near the airport. The point is that places where airplanes or boats depart or arrive flourish.

(Restaurant owner M, Anbo)

Other villages such as Isso and Nagata may be regarded as disadvantaged areas. There were three accommodations located in Isso village, but two of them were closed and one will be closed soon due to lack of tourists as well as the owner getting older; there is no successor:

The area with the largest number of loggerhead turtle eggs is called Nagata area, but Nagata is in the northwest. The weather is the worst in this season [in March]. Nobody goes there even during daytime because it is a peripheral part of the island. That is why the only people who visit there are those who plan to go around the island, and they buy stuff while going around, so they do not spend much money in Nagata.

(Tour guide B, Anbo)

However, other two interviewees attributed that there was not much difference among villages (Tour guide I, Miyanoura). Another tour guide mentioned that the inequality between villages is not a major concern anymore due to the tourists decreasing:

> Well, I thought that big towns were thriving from tourism, but since there are less tourists now, no one says that anymore. When Yakushima was chosen as WHS, the number of tourists reached its peak. There were even local people annoyed by it. But since the Lehman shock, as tourist numbers have decreased, such stories have gone.
>
> (Tour guide C)

Many interviewees pointed out the gap among villages as spatial differences of tourism impacts. These perspectives suggest that the inequality of tourism development still exists on the island. This might affect how local communities benefit from tourism in Yakushima.

5.6   Discussion

Based on the local tourism stakeholders' perspectives, it could be shown that tourism has made a significant contribution to the economy in the research area, having effects on the economy and generating employment for people similar to the research of Franzidis and Yau (2017) on community's perceptions of tourism development. Nevertheless, it also revealed that tourism development has effects on resident turnover as well as disparities in the number of tourism facilities between tourist-villages and non-tourist villages. It is clearly displayed among Miyanoura, Anbo and other villages as the data provide in Chapter 3 show and as perceived by the most interviewees.

Furthermore, respondents also perceive inequalities around income distribution and spatial differences. Hence, the findings suggest that tourism in Yakushima may not necessarily contribute to equal income distribution among local enterprises. The findings of the study also indicate that Yakushima may face similar challenges as Small Island Developing States (SIDS), as highlighted by Scheyvens and Momsen (2008). Although Yakushima is not a sovereign nation, being a small island with a strong focus on tourism, it may encounter comparable issues to those experienced by SIDS. In many small islands, tourism is one of the major industries, however, the growth of tourism in Yakushima perhaps is not synonymous with poverty reduction. In some cases, it would strengthen existing inequalities. These trends in the case of Yakushima may be an example of tourism worsening inequality in Japan. The research findings tend to support previous arguments that tourism contributes to income inequalities among regions and tourism participants (Alam & Paramati, 2016; see also Blake, 2008; Incera & Fernández, 2015; Lee & O'Leary, 2008; Lee, 2009).

While tourism has undeniably contributed to the local economy, there are certain limitations associated with it. First, the seasonal nature of tourism on Yakushima creates a significant disparity between the peak and off-season periods. This fluctuation poses challenges to the stability of tourism-related jobs and the income of local tourism businesses. Second, income distribution is perceived differently among individuals, not only across different sectors but also within the same tourism sector. Varying perceptions of how money flows within the industry can lead to negative attitudes among local residents, particularly if they believe that a substantial portion of the benefits is leaving the island. Third, negative perceptions regarding income distribution can give rise to tensions within local tourism enterprises, social conflicts within communities and other related issues, as discussed in previous studies (Abdullah, Doucouliagos, & Manning, 2015; Lee, 2009). Furthermore, the voices of the Yakushima communities hold significance and can influence tourism development policies on the island. It is essential to consider their perspectives and engage them in decision-making processes related to tourism development.

Unlike other research that applied quantitative methods for measuring tourism impacts and income distribution, the researcher focused on perceptions of local enterprises. This is important for tourism-development policies in Yakushima since this island is facing some issues of overuse, pollution and disparity. It is also a crucial consideration for local authorities in the context of economic development and conservation in an NWHS in the island.

This chapter has presented that local tourism businesses have different perspectives about income distribution among local tourism stakeholders. Accommodation and tour guides were attributed to have more profits than other enterprises, followed by rental car companies and souvenir shops. This point of view suggests that the interviewees tend to see sectors other than their own as more profitable. More than half of the respondents also considered themselves as belonging to a low- and average-income group, even compared within the prefectural level. This evaluation also could be their own choice. As noted, it is true that many people (e.g., guides) who choose Yakushima for their place of living do so as a lifestyle choice. Therefore, income may not be a major factor for their decision to work in tourism. As a result, these people's income cannot be evaluated based on the average living standard. Yet, the perceptions of these people still revealed that economic benefits from tourism are seen as limited and unequal. The self-evaluation of respondents supports the statistics shown about the poverty rate in Kagoshima prefecture. The findings suggest that there is a need to look at perceptions of local tourism businesses and local people at the same time to have a better picture of income distribution. The comparison is necessary to understand the perspectives on tourism development in a certain tourist destination to avoid tensions and inequality between tourism-related groups and non-tourism-related groups on Yakushima.

Reference list

Abdullah, A., Doucouliagos, H., & Manning, E. (2015). Does education reduce income inequality? A meta regression analysis. *Journal of Economic Surveys, 29*(2), 301–316.

Alam, M. S., & Paramati, S. R. (2016). The impact of tourism on income inequality in developing economies: Does Kuznets curve hypothesis exist? *Annals of Tourism Research, 61,* 111–126.

Blake, A. (2008). Tourism and income distribution in East Africa. *International Journal of Tourism Research*, *10*, 511–524.

Franzidis, A & Yau, M. (2017). Exploring the differences in a community's perception of tourists and tourism development. *Tourism Planning & Development*, *15*(4), 382–397

Incera, A. C., & Fernández, M. F. (2015). Tourism and income distribution: Evidence from a developed regional economy. *Tourism Management*, *48*, 11–20.

Lee, S. (2009). Income inequality in tourism services-dependent counties. *Current Issues in Tourism*, *12*(1), 33–45.

Lee, S., & O'Leary, J. T. (2008). Determinants of income inequality in US nonmetropolitan tourism-and recreation-dependent communities. *Journal of Travel Research*, *46*(4), 456–468.

Scheyvens, R., & Momsen, J. H. (2008). Tourism and poverty reduction: Issues for small island states. *Tourism Geographies*, *10*(1), 22–41.

Yakushima Town. (2018a). *Official certified guide of Yakushima*. Retrieved from www. yakushima-eco.com/ (In Japanese).

6 Perspectives of local people on tourism development

The case of Cu Lao Cham

6.1 Introduction

The context of tourism development in Cu Lao Cham sheds light on the present socioeconomic conditions of the island, which are explored in detail in Chapter 4. To address the research question of how the local community perceives the impact of tourism on poverty alleviation, as well as the barriers to tourism participation in the case of Cu Lao Cham, the author conducted semistructured interviews with the locals working in tourism. This chapter presents the findings and analysis of these interviews, which were further supplemented by the author's field notes taken during the study. The chapter commences with a brief description of the interviewees, followed by an examination of the major themes that emerged from the interviews. The local peoples' perspectives on three key themes were identified: (1) tourism's contribution to the locals' livelihoods and the island's economy, (2) the distribution of tourism income and its impact on social conflicts, and (3) the obstacles to tourism participation.

6.2 Interview respondents' demographics

Semistructured interviews, using a nonrandom sampling method, were conducted with local people working in the tourism in all four villages on Cu Lao Cham: Bai Lang, Bai Ong, Thon Cam and Bai Huong in June 2018. The respondents include herbal tea sellers, homestay owners, motorbike taxi drivers, restaurant servers, seafood and souvenir sellers, tour guides, tourist boat operators and beach vendors (hereafter collectively referred to as local tourism enterprises). As shown in Table 6.1, of the 41 interviewees, 24 were female (58.6%) and 17 were male (41.4%). On average, interviewees were 44 years old. Most of them (38) were married, while only three were single. Among those who were married, each had an average of two children. Two respondents had finished college or university, five had only completed high school, 25 had only finished secondary school and six had only completed primary school. Three persons were illiterate (i.e., no school). In terms of residency, 24 interviewees were born on Cu Lao Cham, while 17 migrated from the mainland. Among those who came from outside the island, the average

DOI: 10.4324/9781003496748-10

Table 6.1 Full breakdown of interviewees' demographics

No.	Name pseud-onym	Age	Gender	Married	No. Children	Type of tourism participation	Length of partici-pation (years)	Educational level	House ownership	Monthly average income*	Financial self-evaluation	Village
1	Chia	51	Female	Yes	1	Herbal tea seller	3	Secondary school	House owner	US$ 150	Poor	Bai Ong
2	Man	58	Female	Yes	5	Herbal tea seller	5	Primary school	House owner	Unstable income	Average	Bai Ong
3	Bon	61	Female	Yes	2	Herbal tea seller	5	Illiterate	Given house	US$ 150	Poor	Bai Ong
4	Chi	63	Female	Yes	5	Herbal tea seller	4	Illiterate	House owner	US$ 150	Poor	Bai Huong
5	Hoa	42	Female	Yes	2	Homestay	6	Secondary school	House owner	US$ 300	Average	Bai Ong
6	Quy	57	Male	Yes	2	Homestay	10	Secondary school	House owner	US$ 750	Better off	Bai Lang
7	Suot	51	Female	Yes	2	Homestay	10	Secondary school	House owner	US$ 750	Average	Bai Lang
8	Du	58	Male	Yes	2	Homestay	5	Secondary school	House owner	US$ 300	Average	Bai Huong
9	Manh	57	Male	Yes	4	Homestay	10	Secondary school	House owner	US$ 400	Average	Bai Huong
10	Cung	29	Male	Yes	2	Motorbike taxi	3	High school	No private house	US$ 150	Poor	Bai Lang
11	Bieu	48	Male	Yes	2	Motorbike taxi	4	Primary school	House owner	US$ 300	Average	Bai Lang
12	Tinh	32	Male	Yes	3	Motorbike taxi	4	Secondary school	No private house	US$ 450	Average	Bai Lang
13	Bong	44	Female	Yes	2	Motorbike taxi	2	Secondary school	House owner	US$ 150	Poor	Bai Ong
14	Men	26	Male	Not yet	0	Motorbike taxi	2	Secondary school	No private house	US$ 300	Poor	Bai Ong
15	Hue	31	Female	Yes	3	Motorbike taxi	2	High school	No private house	US$ 150	Poor	Bai Lang
16	Nhat	35	Male	Yes	1	Motorbike taxi	5	Secondary school	No private house	US$ 250	Poor	Bai Lang
17	Sang	26	Male	Yes	1	Motorbike taxi	4	Secondary school	No private house	US$ 300	Average	Bai Lang
18	Tiet	42	Male	Yes	2	Motorbike taxi	5	Secondary school	House owner	US$ 300	Average	Bai Lang
19	Bien	43	Female	Yes	2	Restaurant server	8	Primary school	House owner	US$ 200	Poor	Bai Lang

No.	Name	Age	Sex			Occupation		Education	House	Income*	Status	Place
20	Em	47	Female	Yes	3	Restaurant server	2	Primary school	House owner	US$ 250	Average	Bai Ong
21	Sen	25	Female	Not yet	0	Restaurant server	4	Secondary school	No private house	US$ 250	Average	Bai Ong
22	Thom	44	Female	Yes	2	Restaurant server	3	Secondary school	House owner	US$ 300	Average	Bai Huong
23	Phu	46	Female	Yes	2	Restaurant server	3	Secondary school	House owner	US$ 300	Poor	Bai Huong
24	Lam	33	Female	Yes	1	Seafood seller	3	Secondary school	No private house	US$ 400	Average	Thon Cam
25	Thao	42	Male	Yes	2	Seafood seller	7	Secondary school	House owner	US$ 300	Average	Bai Lang
26	Tien	50	Female	Yes	2	Seafood seller	5	Secondary school	House owner	Unstable income	Average	Thon Cam
27	Giang	38	Female	Yes	2	Seafood seller	11	Secondary school	House owner	Unstable income	Average	Bai Lang
28	Hong	29	Female	Yes	2	Souvenir seller	7	High school	Rental house	US$ 350	Average	Thon Cam
29	Be	26	Female	Yes	2	Tour guide	2	University graduate	No private house	US$ 300	Poor	Bai Lang
30	Hop	33	Male	Not yet	0	Tour guide	3	College graduate	No private house	US$ 350	Average	Bai Lang
31	Giau	30	Female	Yes	2	Tour guide	4	Secondary school	Rental house	Unstable income	Poor	Bai Lang
32	Loc	30	Female	Yes	3	Tour guide	2	High school	House owner	Unstable income	Average	Bai Lang
33	Bau	56	Male	Yes	4	Tourist boat operator	13	Secondary school	House owner	US$ 500	Better off	Bai Ong
34	Quyet	64	Male	Yes	3	Tourist boat operator	7	Primary school	House owner	US$ 500	Better off	Bai Ong
35	Duong	50	Male	Yes	2	Tourist boat operator	7	Secondary school	House owner	US$ 500	Average	Bai Lang
36	Nghia	61	Male	Yes	3	Tourist boat operator	9	Secondary school	House owner	US$ 500	Better off	Bai Lang
37	Tung	56	Male	Yes	3	Tourist boat operator	4	Secondary school	House owner	US$ 300	Poor	Bai Huong
38	Lan	65	Female	Yes	1	Vendor	2	Illiterate	House owner	US$ 50	Poor	Thon Cam
39	Thuyen	42	Female	Yes	2	Vendor	5	High school	No private house	US$ 200	Poor	Bai Lang
40	Thich	41	Female	Yes	3	Vendor	2	Secondary school	No private house	US$ 250	Poor	Bai Lang
41	Linh	37	Female	Yes	2	Vendor	2	Primary school	Given house	Unstable income	Poor	Bai Ong

Source: * Monthly average income is stated in tourism season.

length of residency was 34 years. Twenty interviewees lived in Bai Lang, 11 in Bai Ong, six in Bai Huong and four in Thon Cam.

Most respondents had multiple jobs, which mainly comprised fishing and tourism services. Regarding the length of tourism service participation, the longest participation time was 13 years, the shortest was two years and the average was five years. In terms of income, only one respondent earned less than US$1.90 per day. However, 17 interviewees still considered themselves as poor when compared to the living conditions in urban areas or with people in their neighborhoods. Twenty-six respondents (60.3%) owned a house (i.e., a typical detached house), 11 (26.8%) had no private house (typically sharing a house with their parents), two (4.9%) lived in a rental house and two (4.9%) were offered a house by the local authority due to their poverty. The next section explores the contribution of tourism to respondents' lives and the local economy on Cu Lao Cham.

6.3   Tourism's contribution to respondents' lives and the local economy

When asked about the contribution of tourism to their lives, respondents tended to mention financial aspects (see Table 6.2, which summarizes this section's findings). Over half (28) of respondents said that their living conditions had been improved significantly due to tourism. For instance, Quy, a homestay owner, stated:

> My living conditions have improved considerably since I began operating a tourism service. Previously, I went fishing about 25 days per month but earned only around VND2,000,000 to VND3,000,000 (US$100 to US$150). There were days when I earned just enough money to pay the cost of diesel fuel. I had no savings. I think tourism has strongly enhanced the lives of people on this island, including myself.

Quy can be considered as one of the successful examples of poverty alleviation based on tourism. Although he was poor before, he was now very satisfied with his

Table 6.2 Summary of perceived impacts of tourism on respondents' lives and the local economy

Perceived impacts of tourism	N	%	Group of respondents*
On respondents' quality of life			
Improved a lot	28	68.3	Homestay, tourist boat operator, motorbike taxi, tour guide
Improved a little	6	14.7	Tour guide, restaurant server
No change	7	17.0	Homestay, vendor, herbal tea seller
On the local economy			**Group of benefits***
Providing jobs for the poor	37	90.2	Vendor, herbal tea seller, motorbike taxi
Providing jobs for women	37	90.2	Restaurant server, vendor

Source: *Multiple responses are calculated

income and, at the time of the interview, was able to save about VND60,000,000 (US$ 3000) per year. Quy not only ran a homestay as his main business but also operated a tourist boat service. Recently, he had bought two motorbikes to rent to tourists.

Similar to Quy, Bau and Nghia, who ran a tourist boat service, stated that their lives had also have been enhanced due to their involvement in tourism. Previously, their fishing job was difficult and dangerous due to regular natural disasters. At the time of the study, they worked in the tourism sector, which they considered to be easier, with more free time and higher income. They organized fishing trips for tourists spanning a relatively short time (four to five hours), earning Bau and Nghia about VND500,000–VND1,000,000 (US$25–US$50). This income was much higher than their previous fishing job. With his new income from tourism, Bau said that he purchased some high-tech equipment for his family such as an air-conditioner, a refrigerator, and a bathroom equipped with a hot-cold water system. Meanwhile, after three years working in tourism, Nghia repaired his house, replacing his cement floor with an enameled tile floor. He also built two more rooms to rent to tourists.

When compared to commercial fishing or other work respondents were engaged in previously, tourism brought a better income, as Hong, a souvenir seller, said:

> My life has become better since participating in tourism. I worked for a restaurant in Hoi An before with a low wage. It was about VND3,000,000 (US$150) per month, but my living expenses were about VND2,500,000 (US$125), and I saved only VND500,000 (US$25). Now I can earn about VND8,000,000 (US$400) per month in the tourism season, and I can save about VND2,000,000 (US$100). Living expenses and shopping expenditures are also more comfortable than before.

Similarly, Duong, a tourist boat operator, said that his life had changed positively since his time in the fishing industry. His meals were now affordable, and it was easy for him to make ends meet. Previously, his standard of living was poor and sometimes he had to borrow money from his friends or relatives. Compared to fishing activities that brought unstable income, tourism service offered easier and more stable income. Since becoming involved in the tourism sector, he bought a motorbike and upgraded to a bigger television. Meanwhile, Sang, a motorbike taxi driver, related that:

> Prior to involving in tourism, my living condition was poor. I had no refrigerator, washing machine, or other necessary things. Now I earn money from driving a motorbike taxi for tourists. I bought a television, refrigerator and washing machine. Since I purchased these devices for my family, life has gotten a lot better.

Unlike Quy, Bau and Nghia, some other respondents, such as Manh and Du, who lived and ran their homestay service in Bai Huong, have faced some problems.

They wanted to improve their living condition and so became involved in tourism, but there had been no guests since they began operating the services, which they claimed was mainly due to two reasons. First, Bai Huong is located quite far from the main tourist spots (which are mostly located in Bai Lang) and if tourists want to stay overnight, they often prefer to stay at Bai Lang village. Second, although there have been some guests spending several nights in Bai Huong, they gravitate toward homestays that either can be booked and searched on the internet or to which they have been introduced by previous guests. As a result, the living conditions for Manh and Du had not improved at all, and they still relied on fishing for their income.

Although tourism had significantly enhanced most respondents' living conditions, there were some respondents, especially for those who participated in unskilled and/or less-skilled tourism-related jobs such as herbal tea sellers and vendors on the island, whose lives had not changed much since working in tourism. They still considered themselves as poor and low-income earners, shown in Table 6.1. For example:

I am not satisfied with my current income. This income is not enough for my family's living expenses, including my children's schooling. The school expenses for high school in the mainland [high school is not available on the island] are high. I am worried about how to improve my income so that I have more money to support my children's study. I want them to study as high as possible to have a better job, not work hard as a manual vendor like me.

(Thuyen, vendor)

Overall, tourism had meaningfully enriched the respondents' lives. Most interviewees used their earnings from tourism to pay for daily necessities (e.g., food and clothing) and to purchase more expensive items such as motorbikes, televisions, refrigerators and washing machines. For others, the financial gain from tourism was reinvested in long-term strategies to earn even more money, for example, by purchasing more motorbikes to rent to tourists (Quy) or building more rooms to rent to guests (Nghia). Some respondents used their earnings to support their children's education, with hopes of improving their children's lives in the future (Thuyen).

When asked about the contribution of tourism to the local economy, particularly in employment generation, over half (26) of respondents were satisfied with the current tourism development on the island. They stated that tourism had created many jobs for local residents, including poor people and women (see also Table 6.1). The island's poor people had the opportunity to become sellers of food, drinks or herbal tea. Meanwhile, women were frequently involved in restaurant services. In Vietnamese culture, especially in the fishing industry, women are often not allowed to take part in certain activities because of traditional beliefs (i.e., a woman's menstrual cycle may affect the sea environment) and longstanding divisions of labor by gender. Because of these taboos and customs, women typically stayed at home, carrying out domestic responsibilities, and had no jobs. Tourism had unlocked opportunities for women to engage in tourism services in Cu Lao Cham MPA, as Nghia said:

> Before, my partner was a housewife. Now she also takes part in tourism, such as taking care of guests and cooking for them, to earn more income. My neighbors also switched from fishing to tourism services due to the low productivity of fishing. Some of them became food sellers, and others participated as motorbike taxi drivers.

Similarly, Sen, a female restaurant server, and Be, a female tour guide, indicated that tourism created more jobs for women. Both spoke about how, before the development of tourism, women relied on their husbands for income, or went out to collect firewood in the forest to sell. Now, they worked at restaurants or sold food, drinks or souvenirs to tourists. Overall, the advent of a tourism economy on the island generated more employment opportunities, especially for women and young people.

With respect to tourism's contribution to generating employment for poor people, aside from the aforementioned jobs, more poor people were also involved in driving motorbike taxis for tourists. To facilitate local people's participation in tourism services, the THPC has established two motorbike taxi groups. Each group has about 38 driving members. In order to become a member of these two groups, a person needs to be classified as poor. This initiative was an efficient model for helping poor participate in tourism and facilitate their income. Without such schemes, they may have no work available, as related by Hong:

> Tourism contributes about 90% of job generation for local people. Previously, people mainly went fishing and collecting firewood. Because marine resources are not as bountiful as before, today many people have become motorbike taxi drivers to have a better and more stable income. They can earn about VND100,000 to VND200,000 (US$5 to US$10) per day if they have customers.

In general, in the off season for tourism, some respondents may return to fishing to make a living, but most of them have nothing to do. Despite the seasonal effect of tourism on the island, tourism is crucial to the island's economy, not only in creating more jobs for local people but also in allowing them to maintain their lives economically throughout the year. Having said that, many informants stated that they had to save money to make up for shortages in the off-season. The majority of respondents indicated that, without tourism development on the island, local people would face significant difficulties and that many would have continued to live in severe poverty.

6.4   Tourism income distribution and social conflicts

Despite the positive contributions of tourism to respondents' incomes and to the local economy, respondents also expressed negative perceptions of the impact of tourism, particularly regarding unequal income distribution and social conflicts (refer Table 6.2, which summarizes this section's findings).

According to some respondents, there are problems with the unequal distribution of the economic benefits from the new tourism economy among the individual tourism stakeholders on the island and between the local service providers and tour operators on the mainland. Specifically, for local tourism stakeholders, the restaurant sector was regarded as the main beneficiary of tourism. Meanwhile, islanders working in tourism perceived the mainland tour operators to be receiving greater economic benefits than on-island tourism providers. For example, Hong, a souvenir seller, said:

> The restaurant sector on the island has more opportunities to benefit [from tourism] because restaurants have customers continuously [during the tourist season]. Restaurants not only serve food, but also sell other things like swimsuits and seafood at prices much higher than is typical for the area. Because of this, restaurant owners earn very high income. Meanwhile, tour operators on the mainland have high profits because they sell one-day tours at quite a high price, about VND650,000 (US$32.50), with or without a meal. The people on the island are limited to small-scale businesses.

Similarly, Duong, a tourist boat operator, and Sang, a motorbike taxi driver, stated that the restaurant sector was the most profitable group among local tourism stakeholders, because they benefited from the package tours sold by tour operators in the mainland. When operators brought tourists to the island on all-day tours, the restaurants were guaranteed customers. Local tourism stakeholders on the island needed to pay operational costs similar to the restaurants, but they did not have customers constantly. Meanwhile, Duong stated that although tourism created more jobs for local people, it brought more benefits for tour operators in the mainland. One of the reasons was that most of the tourists booked a limited package tour to Cu Lao Cham, coming and going within a single day's span. The use of the island's tourism services and tourists' spending in general on the island was minimal, and therefore the direct benefits for local people were also limited. Sang estimated that tour operators earned 80% of what tourists spent, because they could attract large groups of tourists and sell tours at a high price. This can be viewed as economic leakage of tourism benefits on Cu Lao Cham, as most tour operators are owned by tourism enterprises on the mainland.

Opinions among groups were varied, however, with several interviewees (such as Tien, a seafood seller; and Suot, a homestay owner) relating that motorbike taxi groups in fact received more benefits from tourism because they did not need to pay taxes and could earn income as long as they had customers who needed transportation, including those on day trips. Unlike the motorbike taxi drivers, homestay owners made money only if tourists spent the night on the island. Meanwhile, Be, a tour guide, said that because tourism development on Cu Lao Cham was community-focused, it was not directed at individual beneficiaries, but was intended to benefit everyone. Comparing who earns more or less on Cu Lao Cham indicates that this ideal may not necessarily hold true on the island (Be).

The perceived benefit that the restaurants on the island and the tour operators in the mainland were the most profitable sectors, can be explained in several ways. First, in order to establish restaurants or large-scale tour operations, large initial costs are required. Low-income or poor individuals are, therefore, precluded from involvement in such businesses due to a lack of financial capital. The more capital is possessed, the more benefits generated may be achieved. Second, on Cu Lao Cham, most people have a low level of education, which may prevent them from participating in skilled and higher-income jobs. Third, as discussed earlier, the main source of restaurants' customers are the tour operators, and thus the more tourists come to the island, the more benefits for the restaurants. Fourth, tour operators play an active role in searching for tourists, organizing tours, and thus it is understandable that they benefit more from tourism. This suggests that the interviewees may tend to see sectors others than their own as more profitable.

These results show that tension existed among various tourism service participants on the island, and between local tourism stakeholders and tour operators in the mainland (see also Table 6.3). As noted, the price of a package tour offered by travel agents/tour operators was perceived to be too expensive and was also increasing. Because of the rising price of the tours, the number of tourists visiting Cu Lao Cham and the expenditures on the island had been decreasing gradually, according to some respondents. For example:

> The price of a one-day tour for one person to Cu Lao Cham was fairly cheap before, about VND450,000 (US$22.50). Now it has increased to VND650,000 (US$32.50). It is not so expensive for a single person, but it is really expensive for a group of people. General speaking, due to the high price of the tour, tourists do not want to spend more money on the island.
>
> (Tinh, motorbike taxi driver)

Table 6.3 Summary of perceived income distribution and social conflicts

Perceived income distribution	*N*	*%*	*Main tourism beneficiaries*
Unequal among local tourism stakeholders	29	72.5	Restaurant sector on the island
Economic leakages	17	41.4	Tour operators in the mainland
Perceived social conflicts			**Group of conflicts**
Between local tourism stakeholders with tour operators	12	29.2	Tourist boat operator, tour operator
Between local freelance guides and the package tour guides	4	0.97	Local guide, package guide, tour operator
Within local tourism stakeholders	5	12.1	Tourist boat operator, seafood seller

In addition, as Duong said, tour operators tended to include some tourism services on Cu Lao Cham within tour packages. In the beginning, tour operators simply organized and sold tours to the island. But operators began to include local services (such as transportation and fishing trips offered by local boat owners). In this way, they provided tourists with local services for less than they would otherwise pay. By assuming this intermediary role, tour operators provided customers to local businesses, but those local businesses would have earned higher profits if they got their business from tourists directly. This phenomenon was a source of disagreement between tour operators and local tourism stakeholders on Cu Lao Cham. Some respondents had ideas for mitigating the aforementioned conflicts. For example, Duong and Suot had similar points:

> I want the tour operators bringing tourists to the island to let [the tourists] have free time. Tourists should be free to go anywhere and buy anything they want. Currently, the tour operators cover all services, and tourists have no time to buy anything. This causes difficulties for local small businesses.
>
> (Duong)

Furthermore, there was a potential source of conflict between local freelance guides on the island and the package tour guides and/or tour operators from the mainland. Loc and Giau, who worked as local guides, may have been more knowledgeable about tourist spots on Cu Lao Cham than the mainland tour guides. Yet increasingly, these local guides only had customers if the tour operators needed some extra guides for the package tours. Loc said that she had a lot of tours and earned a lot of income the year prior. Working 20 days each month, she earned around VND15,000,000 to VND20,000,000 (US$750 to US$1000). However, at the time of interview, she did not have much work, because many tour operators brought their own tour guides for the package tours as a cost-saving measure. Due to their dependence on tour operators on the mainland, many local guides were faced with a dilemma: continue with their current, economically unstable job, or look for another job, despite there being a lack of opportunities. Because of this situation, they expressed hope that a local guide group or association would be established to protect their livelihoods (Loc & Giau).

Social conflicts were also present within the local tourism stakeholders. For instance, Duong described tension among tourist boat operators. Some operators reduced the price of their services to attract customers, but the services they provided were considered to be of inferior quality. Duong described this as being unfair competition. Tensions were also present among sellers of commodities (e.g., seafood and souvenirs), some of whom sold their merchandise at the tourist market, while others set up on the sidewalk or operated out of their homes. Those who sold their commodities at the market had to pay initial costs (about VND20,000,000 (US$1000) for a three-year contract) to reserve a selling space at the tourist market (near the pier where tourists disembark) and also had to pay annual taxes. The merchants operating out of their homes did not need to pay those initial costs. The tourist market was created as the main place for tourists to shop. However, according to Hong and Tien, who were souvenir sellers, recently tourists did not buy much at the market, instead shopping elsewhere due to limited time.

There are no conflicts between the two motorbike taxi groups (with 38 drivers each, as mentioned earlier) due to their organizational structure. According to the interviewees and the author's observations, the method of running this service was organized as follows: one group took a position near the pier serving tourists, while another group approached tourists at the Cu Lao Cham MPA exhibition center (see Figure 6.1). The groups rotated between these positions each day. Members of either group could solicit individual tourists for motorbike rides as well as serve organized tour groups. The drivers took turns transporting customers to ensure that every member had work. Each group had a leader who managed the group and was responsible for collecting payment from the customers. The day's earnings were divided equally among group members. In this way, every driver was guaranteed income as long as their group had customers. Thus, the drivers avoided conflict.

However, motorbike taxi drivers Tinh and Hue did raise concerns that tourism only offered benefits to those who lived near the pier used by tourists (i.e., port-based services), while people living far from the pier did not benefit much from tourism. Issues like this may lead to conflicts among local people, and particularly between tourism-related and non-tourism-related groups on the island. Although, at the time of this study, the conflicts and tensions had just begun, the negative

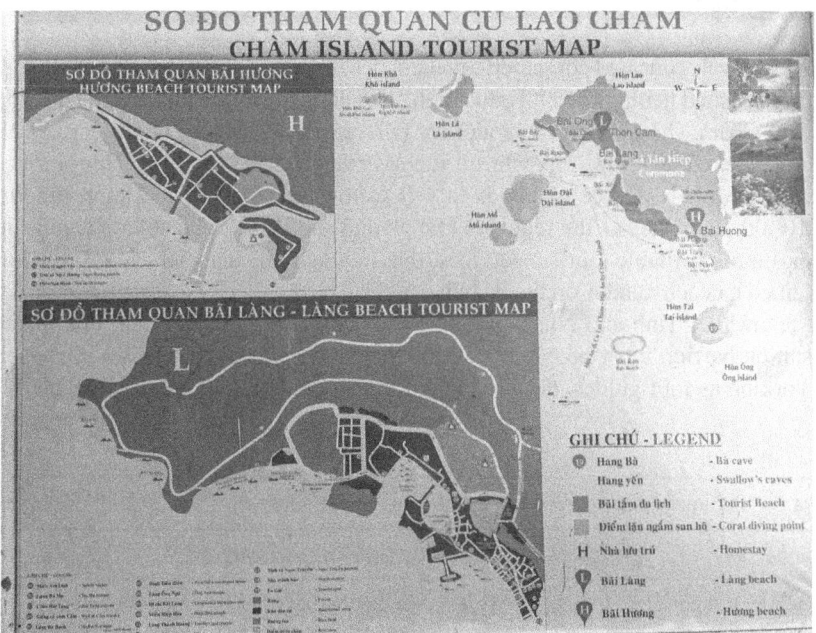

Figure 6.1 Tourist map of Cu Lao Cham.

impacts of tourism development on local communities and conservation efforts in Cu Lao Cham MPA could worsen.

Overall, despite the inequality of income distribution and certain social conflicts occurring on Cu Lao Cham, all interviewees agreed that tourism was the best option available to replace increasingly unviable work in the fishing sector, and they tended to support the continuing development of tourism in the study site. These findings again helped confirm the results discussed from the questionnaire.

6.5   Barriers to tourism participation

The 35 respondents asked about the challenges of tourism services pointed to different obstacles depending on the type of work they performed and their own particular situations (see Table 6.4, which summarizes this section's findings). The barriers were related to personal circumstances (e.g., lack of financial capital, health conditions); market access (e.g., networking issues, location of business); and local policies (e.g., conditions for participating in the motorbike taxi groups). For example, as Thuyen said:

Poor people like me often face barriers because we lack financial capital to participate in tourism services. Indeed, the CLCTMB does not allow me to sell merchandise in some areas. They suggested that I try to secure a selling space in the market so that I can sell dried seafood, but I do not have enough money for that.

Thuyen is well educated compared to others beach vendors on the island. Although she has finished her high school education, a level of education that is quite high on Cu Lao Cham, she works as a vendor for two reasons. First, her family has a fruit garden, where they grow tropical fruit such as mangoes and jackfruit. This produce gives her a ready source of goods to sell to tourists. Second, and more importantly, although she has the requisite knowledge and skills to be able to engage in other, more profitable tourism services, due to a lack of financial capital, she has remained a beach vendor on the island.

Meanwhile, Tinh and Nhat, motorbike taxi drivers, indicated that due to the education barrier, most people could not participate in skilled tourism jobs such as working as tour guides, despite the guide courses being offered by the THPC.

Table 6.4 Summary of perceived barriers to tourism participation

Main barrier	N	%	Group of respondents
Respondents themselves	18	51.4	Vendor, herbal tea seller, motorbike taxi
Market access	14	40	Homestay, tourist boat service, tour guide
Local policies	3	8.6	Motorbike taxi

Meanwhile, the island's restaurants occasionally hired part-time workers, but they did not employ middle-aged or elderly people. Thus, age and health conditions also became obstacles for participation in tourism services. Finally, according to the interviewees, there was no system of grants from NGOs or local government that would provide the island's poor with start-up capital for establishing tourism-related businesses, beyond what is offered by the MPA project, Cu Lao Cham's livelihood support program that was funded by DANIDA.

The networking required to develop a base of customers is an important factor for participating in tourism services and maintaining businesses. For example, as noted earlier, if Loc and Giau, the local freelance tour guides, wanted access to tourist customers, they would need to build their own networks with the tour operators in the mainland. Where before they were able to find customers independently, now their success as tour guides is tied directly to their ability to receive customers from the mainland tour operators.

Similarly, Hong and Tien, who ran their businesses in the market near the tourist pier, also relied very much on the package tourists as their main source of customers. As Hong said:

My most pressing difficulty in selling souvenirs here is that I have fewer customers, because tour guides do not give tourists time to go shopping in the market. Before, tourists had 15 minutes of shopping time in the tour program, but now they do not. If tourists are introduced to other services, such as motorbike taxi or scuba diving tours, the tour guides will receive a commission. In general, my business depends on tour guides directing tourists to me, but I do not have much profit to offer as a commission, so the guides prioritize other services when making their schedules.

Only three respondents indicated that barriers imposed through local policies, in particular the conditions set by the THPC for becoming a motorbike taxi member, were the main factor that prevented poor people from engaging in tourism-related work. However, due to the profit sharing policy of the motorbike taxi groups, if more new members join, the net income would be reduced for each member. Meanwhile, the number of tourists to Cu Lao Cham had decreased recently due to the increased price of organized tours. Consequently, according to Hue, there had been about 200 taxi driver applications recently, but only two new applicants were accepted because of their poverty. The volume of applications was quite high, probably due to the increased shift in employment on the island from the fishing sector to tourism services and/or the observed benefits of tourism-related work. Becoming a motorbike taxi driver was one of the easiest ways to earn money on the island, so long as one owned a motorbike. (In Vietnam, motorbike ownership is common.) That said, although the motorbike taxi service was a good model and had created jobs for poor people when the groups were established, if the access to this type of work is limited, it is not a sustainable model.

These barriers to participation on Cu Lao Cham suggest that the local authority had considered the potential contribution of tourism to local people's lives when establishing the Cu Lao Cham MPA. The authority was also proactive in developing policies connecting tourism to poverty alleviation. Therefore, local policies were not the main barrier to participation in tourism services, but rather the particular circumstances of respondents as well as difficulties obtaining market access. Tourism on Cu Lao Cham relies heavily on the tourist market on the mainland, especially package tours sold by tour operators and travel agents. The proportion of tourists visiting Cu Lao Cham independently is relatively low, which suggests that if tourism on Cu Lao Cham faced a shortage of tourists in the future, that would lead to a worsening in conditions for local people working in tourism services and potentially impacting the conservation goals of Cu Lao Cham MPA.

6.6   Discussion

This chapter adopts a bottom-up perspective to examine the perceptions and experiences of local people in Cu Lao Cham regarding tourism as a means of poverty alleviation. Both the questionnaire survey and interviews results suggest that tourism has made some contributions to poverty alleviation, based on the accounts of respondents. In general, the findings from this chapter reveal that tourism has contributed significantly to many local people's lives and to the local economy. This finding is in line with Soliman (2015) that the adoption of PPT in protected areas in Fayoum, Egypt positively impacts the livelihoods of local people.

In many small islands in Vietnam, especially where MPAs have been established, most local people rely on fishing as their traditional mode of employment. Although fishing is vital to the lives of local people, its benefit is limited. Many coastal communities have small-scale fisheries and are highly dependent on the availability of marine resources; thus, fishing productivity is often low and unstable (Pomeroy, Nguyen, & Ha, 2009). Local living conditions also exhibit many difficulties, such as the lack of national electricity and water supply infrastructures. Therefore, when tourism was developed and promoted to replace the fishing sector in Cu Lao Cham MPA, so long as it could help local people improve their lives, locals tended to view it positively.

The contribution of tourism on Cu Lao Cham to the local economy has also provided a strong example of job generation for women, a marginalized group often excluded from sustainable development projects including tourism (e.g., Spiteri & Nepal, 2006; Tran & Walter, 2014; Walpole & Goodwin, 2001). Tourism on Cu Lao Cham has actually unlocked opportunities for women, helped them gain independence and allowed them to earn necessary income for their families. This finding is in line with the study of Leisher, Beukering, and Scherl (2007) where tourism benefits women economically and in some cases socially.

One of the remarkable contributions of tourism to employment generation and to poverty reduction in the study site was the establishment of the motorbike taxi groups. Although this can be seen as a good initiative, modeling the type of collective income source proposed by Roe and Khanya (2002) that offer an opportunity for the poor to participate in and benefit equally from tourism, the introduction of alternative models may be necessary to offer more economic benefits to the local community.

According to the respondents, tourism reduces poverty in Cu Lao Cham MPA for the following reasons. First, the study site itself has rich cultural and natural resources, the latter having been recognized with the establishment of the MPA and WBR, that attract tourists to the island. Second, local policies tend to encourage the local community to become involved in tourism, thereby securing a better job with more income (QNPPC, 2017). Third, since the MPA project was launched in the island, capacity building and awareness-raising of local residents have been considered and the local people have also gained certain knowledge and skills that enabled them to participate in tourism services. This suggests that the MPA project, together with tourism growth on the island, has enhanced to some extent the non-economic impacts of PPT strategies on local people (Roe & Khanya, 2002), which in turn has contributed to poverty reduction (Leisher et al., 2007). Finally, Cu Lao Cham island belongs to the ancient town of Hoi An, a well-known World Heritage Site in Vietnam, which attracts a significant number of international and domestic tourists annually. The tourist flow is of great benefit to Cu Lao Cham.

Although tourism is an important part of many people's lives on the island, many of the interviewees shift between tourism and fishing work so as to have better income. This finding may suggest that although tourism helps boost the respondents' earnings, it is still insufficient for their livelihoods if they must switch back to fishing in the slow season. However, their involvement in multiple industries could also be by choice. This finding is consistent with the study of Lopes, Pacheco, Clauzet, Silvano, and Begossi (2015), who found that tourism service and fishing activity together provide better income than just one mode of employment.

Proponents of PPT have pointed out that the informal nature of the tourism sector offers poor people and small-scale commodity sellers the opportunity to participate in and benefit from tourism (Ashley, Roe, & Goodwin, 2001; Truong, 2018). Indeed, the accounts of the poor people in the research area, particularly the herbal tea sellers, souvenir sellers or other small-scale vendors, suggested that they have benefited from tourism. Yet compared to other participants, the income of poor people on Cu Lao Cham from tourism is quite low because they often lack skills, knowledge, and capital, which affirms the studies by Holden, Sonne, and Novelli (2011) and Huynh (2011). The respondents' income also depended on the type of tourism services that they performed, tourists' demand for those services, their own capacity and competition in the market (Winter & Kim, 2021).

Despite the positive impact of tourism on many respondents' lives and the local economy, this study uncovered perceived inequality of income distribution and social conflicts. This finding is consistent with the results of studies by Alam and Paramati (2016); Blake, Arbache, Sinclair, and Teles (2008); and Scheyvens and Momsen (2008) that tourism creates income inequality in some cases. As discussed in prior studies (Abdullah, Doucouliagos, & Manning, 2015; Jordahl, 2007), unequal income distribution is one of the major problems of economic growth, causing social conflicts and minimizing social trust within communities. Distribution of the benefits of tourism and the attitudes of local communities in the context of protected areas have been widely discussed as the key barriers to success for conservation (e.g., Brooks, Waylen, & Mulder, 2013; Spiteri & Nepal, 2006; Walpole & Goodwin, 2001), and Cu Lao Cham MPA is no exception.

With respect to the barriers to participation in tourism services, although respondents listed different constraints, this chapter's findings revealed that the lack of professional knowledge, poor health, financial capital and market access were the main issues respondents faced. Market access is stated as one of the critical issues for implementing PPT (Ashley et al., 2001), and it was a particular challenge for many local tourism stakeholders in the study site. Market access plays an important role in enabling local tourism providers to maintain their businesses, especially in many rural protected areas. The lack of financial capital has been documented as a barrier to entrepreneurship for poor people elsewhere in Vietnam, such as in Sapa (Truong, Hall, & Garry, 2014) and Lang Co (Redman, 2009), as well as in other developing countries like Bangladesh (Islam & Carlsen, 2012). This contrasts with Fayoum's protected areas in Egypt (Soliman, 2015), where a lack of tourist facilities, accommodation and food or beverage services prompted the local government, local banks and NGOs to establish microcredit programs to provide local people with the capital needed to start tourism services. A similar microcredit system on Cu Lao Cham may alleviate a significant barrier to entry for the island's poor.

Reference list

Abdullah, A., Doucouliagos, H., & Manning, E. (2015). Does education reduce income inequality? A meta regression analysis. *Journal of Economic Surveys*, *29*(2), 301–316.

Alam, M. S., & Paramati, S. R. (2016). The impact of tourism on income inequality in developing economies: Does Kuznets curve hypothesis exist? *Annals of Tourism Research, 61*, 111–126.

Ashley, C., Roe, D., & Goodwin, H. (2001). *Pro-poor tourism strategies: Making tourism work for the poor*. London: Pro-poor Tourism Partnership.

Blake, A., Arbache, J. S., Sinclair, M. T., & Teles, V. (2008). Tourism and poverty relief. *Annals of Tourism Research, 35*(1), 107–126

Brooks, J., Waylen, A. K., & Mulder, B. M. (2013). Assessing community-based conservation projects: A systematic review and multilevel analysis of attitudinal, behavioral, ecological, and economic outcomes. *Environmental Evidence, 2*(2), 1–34.

Huynh, B. T. (2011). *The Cai Rang floating market, Vietnam: towards pro-poor tourism?* (Doctoral dissertation, Auckland University of Technology).

Holden, A., Sonne, J., & Novelli, M. (2011). Tourism and poverty reduction: An interpretation by the poor of Elmina, Ghana. *Tourism Planning & Development*, *8*(3), 317–334.

Islam, F., & Carlsen, J. (2012). Tourism in rural Bangladesh: Unlocking opportunities for poverty alleviation? *Tourism Recreation Research*, *37*, 37–45.

Jordahl, H. (2007). *Inequality and trust* (IFN Working Paper No. 715). Retrieved from https://ssrn.com/abstract=1012786 or http://dx.doi.org/10.2139/ssrn.1012786

Leisher, C., van Beukering, P., & Scherl, M. L. (2007). *Nature's investment bank: How marine protected areas contribute to poverty reduction*. London: The Nature Conservancy and WWF International.

Lopes, P. F. M., Pacheco, S., Clauzet, M., Silvano, R. A. M., & Begossi, A. (2015). Fisheries, tourism, and marine protected areas: Conflicting or synergistic interactions? *Ecosystem Services*, *16*, 333–340.

Pomeroy, R., Nguyen, T. K. A., & Ha, X. T. (2009). Small-scale marine fisheries policy in Vietnam. *Marine Policy*, *33*, 419–428.

Quang Nam Provincial People's Committee (QNPPC). (2017). *Decision No. 2494/ QĐ-UBND of promulgating the regulation on management and organization of tourism and sports activities in Cu Lao Cham MPA, Hoi An city, Quang Nam province.* Retrieved October 2018, from www.thuvienphapluat.vn

Redman, D. (2009). *Tourism as a poverty alleviation strategy: Opportunities and barrier for creating backward economic linkages in Lang Co, Vietnam* (Master's Thesis, Massey University, Massey).

Roe, D., & Khanya, U. (2002). *Pro-poor tourism: Harnessing the world's largest industry for the world's poor.* London: IIED.

Scheyvens, R., & Momsen, J. H. (2008). Tourism and poverty reduction: Issues for small island states. *Tourism Geographies, 10*(1), 22–41.

Soliman, M. (2015). Pro-poor tourism in protected areas—Opportunities and challenges: "The case of Fayoum, Egypt". *Anatolia—An International Journal of Tourism and Hospitality Research, 26*(1), 61–72.

Spiteri, A., & Nepal, S. (2006). Incentive-based conservation programs in developing countries: A review of some key issues and suggestions for improvements. *Environmental Management, 37*, 1–14.

Tran, L., & Walter, P. (2014). Ecotourism, gender and development in northern Vietnam. *Annals of Tourism Research, 44*, 116–130.

Truong, V. D. (2018). Tourism, poverty alleviation, and the informal economy: The street vendors of Hanoi, Vietnam. *Tourism Recreation Research, 43*(1), 52–67.

Truong, V. D., Hall, C. M., & Garry, T. (2014). Tourism and poverty alleviation: Perceptions and experiences of poor people in Sapa, Vietnam. *Journal of Sustainable Tourism, 22*(7), 1071–1089.

Walpole, M. J., & Goodwin, H. J. (2001). Local attitudes towards conservation and tourism around Komodo National Park, Indonesia. *Environmental Conservation, 28*, 160–166.

Winter, T., & Kim, S. (2021). Exploring the relationship between tourism and poverty using the capability approach. *Journal of Sustainable Tourism, 29*(10), 1655–1673.

7 A comparison of two case studies

7.1 Introduction

After examining the two case studies presented in this book, this chapter aims to compare the results obtained from both primary and secondary data sources, which were gathered in Chapters 1–6. First, the chapter highlights the similarities and differences in terms of the geographical characteristics of the two case studies. Second, the tourism policies of Japan and Vietnam are reviewed to identify any similarities and differences. Third, the perceived impacts of tourism on the local economies of the two case studies are discussed in detail. Fourth, the attitudes toward tourism and income distribution in Yakushima and Cu Lao Cham are presented again. Subsequently, the barriers to tourism participation and poverty alleviation in these two cases are further discussed. Finally, based on the findings presented, a new concept of pro-poor tourism in different contexts is proposed.

7.2 Geographical similarities and differences between two cases

After providing an overview of the two case study areas, this section summarizes the similarities and differences between them in terms of various factors such as geographical characteristics, tourism development, economic situations and regulations of nature conservation areas. These details are presented in Table 7.1.

7.3 Tourism's policies

As previously mentioned, although both Japan's and Vietnam's tourism policies aim to place tourism at the center of economic growth by increasing the number of visitors, they differ in terms of their goals for tourism development. Japan's tourism policies focus on three main points. First, tourism is regarded as a crucial sector that contributes to the national economy while promoting cultural exchange, particularly between Japanese culture and foreign visitors' cultures. Second, since the country aims to become a tourism nation, its policies are more geared toward inbound tourism, which is achieved through various strategies and frameworks. Third, because of the uneven distribution of economic development in Japan,

DOI: 10.4324/9781003496748-11

Table 7.1 Comparing case studies: Yakushima and Cu Lao Cham

Category of comparison	Yakushima	Cu Lao Cham
Geographical characteristics		
• Area	505 km²	15 km²
• Population & village	13,000 with 24 villages	2,500 with 4 villages
• Year of establishment NWHS and MPA	NWHS: 1993	MPA: 2005
• Industries & traditional industry	Agriculture, forestry, fishery, and tourism. Forestry was a traditional industry.	Fishery and tourism. Fishery was a traditional industry
Tourism development		
• Starting year	1970	2005
• Main attractions	Forest and mountain landscapes (e.g., Yakusugi trees)	Marine settings (e.g., beaches, coral reefs, seafood)
• Average visitor arrivals within a 5-year period (2013–2017)	284,372 tourists	313,210 tourists
• Tourist activities	Ecotourism activities (e.g., hiking, climbing)	Marine activities (e.g., swimming, diving)
• Average length of stay	2–3 days trip	Daily trip
• Role of tourism sector	Main sector	Main sector
• Tourism season	March–November	March–August
Economic situation	Together with forestry and fishery, tourism plays an important role. However, it has created an inequality of tourism benefits among villages & between locals and non-locals	Tourism has helped local people gain income for a living and reduced poverty. At the same time, it began creating an inequality of tourism benefits
Regulations of nature conservation areas	Regulations are specified in accordance with each type of nature protected systems of "Wilderness Area", "National Park", "Forest Ecosystem Reserve", and "Natural Monuments"	Regulations are specified in accordance with zoning system and its restriction of the MPA
Similarities	• Both cases are nature conservation areas • Tourism is the main industry today • Tourism replaces traditional jobs • Local people are involved in tourism • Both are remote islands and tourism is seasonal	

tourism is seen as a tool to address economic disparities between the developed central regions and other less-favored areas.

Vietnam's tourism policies also have distinct features. In the initial phase of its tourism development (1960–1975), tourism was mainly developed for political purposes due to the Vietnam War's impact, and hence, no tourism policies were formulated during this period. The Vietnamese government began to consider tourism as an economic sector in the next phase of tourism development (1976–1990), and since 1991, various tourism policies linked to poverty reduction have been implemented. Vietnam's tourism development and policies aim not only to stimulate economic growth through tourism but also to alleviate poverty, considering the sector an essential means of achieving this goal.

These findings indicate that Vietnam, on the one hand, is a developing country that is relatively poor in absolute poverty terms, as noted earlier (GOV, 2003; Huxford, 2010). As a result, tourism policies in Vietnam are likely to be more closely linked to poverty alleviation to some extent. On the other hand, Japan is a developed country that also faces income inequality (relative poverty), particularly among regions and people. Although it is recognized that Japan's tourism policies also aim to reduce regional economic disparities, there is a need to pay greater attention to addressing the income gap among tourism stakeholders, as discussed in this study's case of Yakushima.

7.4   Perceptions of tourism's contribution

Based on the interview results presented in Chapters 5 and 6, perceptions of tourism's contribution in both case studies are relatively consistent regarding its impact on the local economies and the lives of local people. Tourism is generally seen as a crucial driver of economic growth and poverty alleviation, particularly in the case of Vietnam. The positive perceived impacts of tourism on the two local economies are significant because tourism is being promoted as an alternative livelihood and is replacing traditional industries for conservation purposes in both cases, as previously discussed (forestry in the case of Yakushima and fishery in the case of Cu Lao Cham). Therefore, local people's positive attitudes toward tourism may help reinforce their support for tourism development and conservation goals simultaneously.

However, the study findings also reveal several differences between the two study sites. In the case of Yakushima, tourism not only creates jobs for local people but also attracts new residents from outside the island who come to work in the tourism sector. Tourism development and the island's natural beauty also attract retirees who settle on the island. However, one issue with tourism's impact on Yakushima is the inequality of employment generation within the tourism sector, with most jobs being created for tour guides, particularly mountain guides, who are mainly recruited from outside Yakushima. This perceived impact may be due to the nature of tourism activities on Yakushima, where most tourists are interested in hiking and exploring the island's mountainous areas, as well as the lack of human resources on the island.

In the case of Cu Lao Cham, although tourism has been developed recently, it has had a significant impact on many local people's lives and the local economy. The findings reveal that tourism has helped lift a number of poor people out of poverty. Tourism has become an important sector on Cu Lao Cham, creating many jobs for poor people, especially women who were previously dependent on their husbands' income. However, the extent to which local people's lives have improved varies depending on the type of tourism services they provide, the level of demand for those services by tourists and the level of competition in the market. This has also led to the issue of inequality of tourism benefits among tourism participants on the island.

Cu Lao Cham has benefited from tourism as it has helped alleviate poverty for many locals. However, in Yakushima, tourism has not been seen as an effective way to reduce income inequality on the island. As a result, the impact of tourism on the two local economies is different, indicating the need for diverse tourism policies to address the impacts of tourism on conservation areas.

The next section re-examines the perceptions of tourism and income distribution in the two case studies.

7.5   Perceptions of tourism and income distribution

The perceived income distribution of tourism in the two case studies shows both similarities and differences, which can be attributed to the unique characteristics of each study site and the varying perceptions of the locals. The study findings indicate that respondents in both cases perceived inequality in income distribution among local tourism enterprises. This suggests that the benefits of tourism are not evenly distributed and depend on various factors such as the type of tourism businesses, demand from tourists and competition in the tourism market. Furthermore, it highlights that people tend to view sectors other than their own as more profitable. Another common perception in both case studies was that tourism profits tend to leak out of the islands rather than benefiting the local population. This negative perception of the benefits of tourism could potentially impact the attitudes of locals toward tourism development and conservation efforts in both cases, as discussed earlier.

Regarding differences in tourism income distribution, the study found that Yakushima's accommodations and tour guides were perceived to have more profits than other enterprises due to the nature of tourism activities on the island. In contrast, in Cu Lao Cham, restaurant owners were considered to benefit the most from tourism, while mainland tour operators were perceived to have greater benefits than tourism services on the island. These perceived differences can be explained by several factors, such as the ownership of the businesses, the main source of customers, and the active role played by tour operators in bringing tourists to the island. Such perceived differences highlight the importance of understanding the unique characteristics of each destination and developing policies that address the specific impacts of tourism on income distribution. Additionally, the study revealed that both case studies shared the perception that tourism profits leak out of the local economy, which may negatively affect local attitudes toward tourism development and conservation efforts.

The case of Cu Lao Cham also saw instances of tension and social conflicts arising between different types of local tourism enterprises, as well as between these enterprises and tour operators on the mainland. These conflicts were fueled by the perceived unequal distribution of benefits among tourism service providers on the island.

Although tourism has recently been developed on Cu Lao Cham, there have been some social conflicts in this case study. Meanwhile, in the case of Yakushima, tourism has been developed for half a century, and the issues of income inequality and spatial disparities have been identified as the most significant challenges. Unequal income distribution is a major problem in economic growth that leads to social conflicts and reduces social trust within communities, as previously discussed (Abdullah, Doucouliagos, & Manning, 2015; Jordahl, 2007). Furthermore, the distribution of tourism benefits and the attitudes of local communities toward conservation in natural areas have been widely discussed as key barriers to conservation success (e.g., Brooks, Waylen, & Mulder, 2013; Spiteri & Nepal, 2006; Walpole & Goodwin, 2001), which is also the case for Yakushima and Cu Lao Cham.

Overall, despite the issue of income inequality, the residents of both Yakushima and Cu Lao Cham showed support for the ongoing development of tourism in their respective study sites.

7.6   Barriers to tourism participation and poverty alleviation

This section will revisit the barriers to tourism participation and poverty alleviation, with a focus on the case of Vietnam. It is important to note that due to the different approaches to measuring poverty in Japan and Vietnam, it was not suitable to examine the perceptions of people regarding barriers to tourism participation and poverty alleviation in the case of Yakushima. However, this issue will still be discussed based on secondary data presented in previous chapters.

As researchers such as Tachibanaki (2006) have noted, income inequality (relative poverty) in Japan can be attributed to several factors at a national level. First, the shift from seniority-based wage payment to performance-based payment has resulted in an income gap between productive and less productive workers. Second, an aging population with a higher proportion of elderly citizens and fewer young people has created income disparities between these two groups. Third, changes in family structure have widened household income disparities due to a decrease in family size and an increase in single-person households. Fourth, Japanese society transitioned toward a free market mechanism that emphasizes competition and discourages government intervention and regulation, thereby resulting in increased income inequality.

At a local level, such as on Yakushima Island, income inequality is likely caused by similar factors. Despite tourism being a significant industry on the island, its development has not necessarily contributed to reducing relative poverty. In reality, tourism growth has widened income gaps between tourist and non-tourist villages on the island (as discussed in Chapter 5), creating a disparity among villagers. This does not imply that the tourism sector itself is a barrier to poverty alleviation on Yakushima, but its unbalanced growth and negative impacts could prevent the mitigation of relative poverty.

While there may be other contributing factors, the factors mentioned earlier are likely the primary barriers to poverty alleviation in the case of Yakushima.

Tourism is considered an important means of poverty alleviation in the case of Cu Lao Cham. Therefore, increasing the capacities and opportunities for local people to participate in tourism services can help reduce poverty. According to the results provided in Chapter 6, although the people stated various challenges to tourism participation and poverty alleviation, it was revealed that the lack of professional knowledge, poor health, financial capital and market access were the main issues respondents faced. As stated by Ashley, Roe, and Goodwin (2001), market access is one of the most significant challenges for implementing PPT. Indeed, this was a critical barrier for many local tourism stakeholders in Cu Lao Cham. As mentioned earlier, tourism on this island heavily relies on the tourist market on the mainland, particularly package tours sold by tour operators and travel agents. This suggests that tour operators and travel agents played a crucial role in attracting tourists to the island, as the livelihoods of many local people heavily rely on the tourism sector and related services. Without the contribution of tour operators and travel agents, the living conditions of the local population would not have improved.

Market access appears to be an important challenge on Cu Lao Cham; thus, networking between local tourism services and travel agents/tour operators should be prioritized through new contracts or regulations that promote long-term mutual benefits. Furthermore, the lack of financial capital has also been documented as a barrier to entrepreneurship for poor people elsewhere in Vietnam, such as in Sapa (Truong, Hall, & Garry, 2014) and Lang Co (Redman, 2009), as well as in other developing countries like Bangladesh (Islam & Carlsen, 2012). This suggests that local governments, local banks, and NGOs may consider establishing microcredit programs to provide local people with the capital needed to start tourism services.

Once the critical barriers to tourism participation have been addressed, poverty alleviation through tourism is more likely to be achieved. Finally, it may be necessary to establish a local tourism association on Cu Lao Cham to enhance and protect local benefits while preventing tension and conflicts.

7.7 Creating a new concept of pro-poor tourism

Looking back at the concept of PPT, which has been discussed in several studies (e.g., Ashley et al., 2001; Holden, 2013; Roe & Khanya, 2002), it is defined as "tourism that generates net benefits for the poor. Benefits may be economic, but they may also be social, environmental or cultural" (Ashley et al., 2001, p. 2). However, the definition does not address the relative distribution of the tourism benefits. Hence, as long as the poor gain net benefits, tourism can still be considered "pro-poor", even if richer people benefit more than poorer people. This suggests that the PPT pays little attention to distributional issues. In some cases, the poorest people may not benefit from tourism, while those in better conditions earn more income (Blake, Arbache, Sinclair, & Teles, 2008; Deller, 2010). Moreover, the PPT does not consider the different types of poverty concepts and measures in different contexts where tourism development may affect poverty situations differently.

Based on the findings of this research, it is evident that tourism can contribute to poverty alleviation and be classified as "pro-poor" tourism, particularly in developing countries like Vietnam. The economic benefits of PPT were clearly reflected in the case of Cu Lao Cham, where tourism also created opportunities for women to participate in the industry. However, it also created inequality in the distribution of tourism benefits. In contrast, in developed nations like Japan, tourism growth may have a negative impact on income inequality, as observed in the case of Yakushima. Nevertheless, tourism growth on Yakushima has attracted people who value the island's lifestyle, indicating that tourism can contribute to the cultural aspect of PPT. Overall, the concept of PPT needs to consider distributional issues and various poverty concepts and measures in different contexts.

Having said that, although the findings of this research are consistent with the concept of PPT in some aspects, three additional features can be added to its definition and approach. First, the PPT concept should also be concerned about equal income distribution among tourism stakeholders, not just generating "net benefits" for the poor. Second, the PPT can be widely discussed wherever tourism is developed, in both developing and developed economies, as poverty is a global issue. Finally, the role of perception in promoting PPT is very important, in the sense that the poor should be given a chance to voice their opinions regarding tourism and poverty alleviation.

Reference list

Abdullah, A., Doucouliagos, H., & Manning, E. (2015). Does education reduce income inequality? A meta regression analysis. *Journal of Economic Surveys*, *29*(2), 301–316.

Ashley, C., Roe, D., & Goodwin, H. (2001). *Pro-poor tourism strategies: Making tourism work for the poor*. London: Pro-poor Tourism Partnership.

Blake, A., Arbache, J. S., Sinclair, M. T., & Teles, V. (2008). Tourism and poverty relief. *Annals of Tourism Research, 35*(1), 107–126.

Brooks, J., Waylen, A. K., & Mulder, B. M. (2013). Assessing community-based conservation projects: A systematic review and multilevel analysis of attitudinal, behavioral, ecological, and economic outcomes. *Environmental Evidence*, *2*(2), 1–34.

Deller, S. (2010). Rural poverty, tourism and spatial heterogeneity. *Annals of Tourism Research*, *6*(1), 36–48.

Government of Vietnam (GOV). (2003). *Comprehensive poverty reduction and growth strategy* (CPRGS). Hanoi: Cartography Publishers.

Holden, A. (2013). *Tourism, poverty and development*. London: Routledge.

Huxford, K. M. L. (2010). *Tracing tourism translations: Opening the black box of development assistance in community-based tourism in Vietnam* (Master's Thesis, University of Canterbury, Canterbury). Retrieved May 2019, from http://ir.canterbury.ac.nz.

Islam, F., & Carlsen, J. (2012). Tourism in rural Bangladesh: Unlocking opportunities for poverty alleviation? *Tourism Recreation Research*, *37*, 37–45.

Jordahl, H. (2007). *Inequality and trust* (IFN Working Paper No. 715). Retrieved from https://ssrn.com/abstract=1012786 or http://dx.doi.org/10.2139/ssrn.1012786

Redman, D. (2009). *Tourism as a poverty alleviation strategy: Opportunities and barrier for creating backward economic linkages in Lang Co, Vietnam* (Master's Thesis, Massey University, Massey).

Roe, D., & Khanya, U. (2002). *Pro-poor tourism: Harnessing the world's largest industry for the world's poor*. London: IIED.

Spiteri, A., & Nepal, S. (2006). Incentive-based conservation programs in developing countries: A review of some key issues and suggestions for improvements. *Environmental Management, 37*, 1–14.

Tachibanaki, T. (2006). Inequality and poverty in Japan. *The Japanese Economic Review, 57*(1), 1–27.

Truong, V. D., Hall, C. M., & Garry, T. (2014). Tourism and poverty alleviation: Perceptions and experiences of poor people in Sapa, Vietnam. *Journal of Sustainable Tourism, 22*(7), 1071–1089.

Walpole, M. J., & Goodwin, H. J. (2001). Local attitudes towards conservation and tourism around Komodo National Park, Indonesia. *Environmental Conservation, 28*, 160–166.

Conclusion

By reading this book, we can draw several conclusions about the tourism sector in Cu Lao Cham. First, it plays a crucial role in the island's economic growth and in reducing poverty. Since the establishment of the MPA, tourism has become an essential alternative livelihood for many locals. However, the benefits of tourism have not been evenly distributed among all the local people. For instance, those who sell herbal tea or work as beach vendors have received only limited benefits from tourism. In addition, tensions and social conflicts have occurred due to the unequal distribution of tourism benefits among local tourism enterprises.

Several critical barriers to tourism participation have also been identified in this study site, such as the lack of knowledge and skills, financial capital and market access. These challenges have hindered sustainable tourism development in Cu Lao Cham. Furthermore, although tourism has improved the lives of many locals, it has its limitations in terms of income generation due to the seasonal effect of tourism and the heavy dependence on tour operators as the primary source of tourists. As a result, some people still rely on fishery as their secondary source of income besides tourism-related jobs.

Therefore, it is argued that while tourism development in Cu Lao Cham can be pro-poor, a significant number of people may return to poverty unless long-term tourism strategies and plans aimed at equal income distribution, mitigating barriers, and enhancing sustainable benefits are carefully considered by the local tourism stakeholders, with an active role played by the local government.

In the case of Yakushima, tourism has contributed significantly to the local economy by generating employment for both locals and non-locals. However, there are limitations to the benefits of tourism, similar to the case of Cu Lao Cham. The seasonal nature of tourism has a significant impact on Yakushima, creating a huge gap between peak and off-season periods that affects the income stability of the people on the island. In addition, differences in the perceived income distribution among local tourism enterprises and the negative impact of tourism on spatial development in Yakushima may create tensions and social conflicts, as found in the case of Cu Lao Cham. These issues may also affect tourism development and conservation policies, as there is a strong relationship between these two development purposes.

DOI: 10.4324/9781003496748-12

Therefore, it is crucial to pay more attention to tourism planning and policies in this study area. By addressing these issues, tourism development can be more sustainable and beneficial for all the people involved in the industry.

This study once again confirms the importance of considering the perceptions of local people not only regarding tourism and poverty alleviation, as supported by previous research (e.g., Akyeampong, 2011; Holden et al., 2011; Islam & Carlsen, 2012; Truong, Hall, & Garry, 2014; Truong, Liu, & Pham, 2020), but also in relation to tourism and conservation goals, as discussed earlier (e.g., Brooks, Waylen, & Mulder, 2013; Spiteri & Nepal, 2006). Once the insights and attitudes of local people toward tourism development in nature conservation areas are deeply understood and tended to, the goals of sustainable tourism development in these special settings are more likely to be achieved.

Reference list

Akyeampong, O. A. (2011). Pro-poor tourism: Residents' expectations, experiences and perceptions in the Kakum National Park area of Ghana. *Journal of Sustainable Tourism, 19*(2), 197–213.

Brooks, J., Waylen, A. K., & Mulder, B. M. (2013). Assessing community-based conservation projects: A systematic review and multilevel analysis of attitudinal, behavioral, ecological, and economic outcomes. *Environmental Evidence, 2*(2), 1–34.

Holden, A., Sonne, J., & Novelli, M. (2011). Tourism and poverty reduction: An interpretation by the poor of Elmina, Ghana. *Tourism Planning & Development, 8*(3), 317–334.

Islam, F., & Carlsen, J. (2012). Tourism in rural Bangladesh: Unlocking opportunities for poverty alleviation? *Tourism Recreation Research, 37*, 37–45.

Spiteri, A., & Nepal, S. (2006). Incentive-based conservation programs in developing countries: A review of some key issues and suggestions for improvements. *Environmental Management, 37*, 1–14.

Truong, V. D., Hall, C. M., & Garry, T. (2014). Tourism and poverty alleviation: Perceptions and experiences of poor people in Sapa, Vietnam. *Journal of Sustainable Tourism, 22*(7), 1071–1089.

Truong, V. D., Liu, X., & Pham, Q. (2020). To be or not to be formal? Rickshaw drivers' perspectives on tourism and poverty. *Journal of Sustainable Tourism, 28*(1), 33–50.

Index